M000201775

SCIENCE AND MYTH

With a Response to Stephen Hawking's
The Grand Design

BY THE SAME AUTHOR

Cosmos and Transcendence:
Breaking Through the Barrier of Scientistic Belief

Theistic Evolution: The Teilhardian Heresy

The Quantum Enigma: Finding the Hidden Key

The Wisdom of Ancient Cosmology:
Contemporary Science in Light of Tradition

Sagesse de la Cosmologie Ancienne

Christian Gnosis:
From Saint Paul to Meister Eckhart

De la Physique à la Science-Fiction

Wolfgang Smith

SCIENCE AND MYTH

With a Response to Stephen Hawking's

The Grand Design

ANGELICO PRESS
SOPHIA PERENNIS

First published in the USA
by Sophia Perennis, 2010
Second Edition © Wolfgang Smith
Angelico Press / Sophia Perennis edition, 2012
All rights reserved

Series editor, James R. Wetmore

For information, address:
Angelico Press / Sophia Perennis, 4619 Slayden Rd., NE
Tacoma, WA 98422
angelicopress.com
sophiaperennis.com

Library of Congress Cataloging-in-Publication Data

Smith, Wolfgang, 1930–
Science and myth: with a response to Stephen Hawking's
The Grand Design

p. cm.

Includes bibliographical references and index.
ISBN 978-1-59731-135-9 (pbk: alk. paper)
ISBN 978-1-59731-136-6 (hardback: alk. paper)
1. Religion and science. 2. Myth I. Title.
BL241.S685 2010
201'.65—dc22   2010011744

Jacket Design: Michael Schrauzer
Jacket image credit: Jacob Heinrich Elbfas,
detail of *Vädersolstavlan*, 1636.

In Memoriam

WERNER PETER SCHMITZ-HILLE

† 24 December, 2008

CONTENTS

PREFACE
TO SECOND EDITION

SOME TWO YEARS AGO, in 2010 to be exact, Stephen Hawking's dazzling and in a way epochal treatise, *The Grand Design*, hit bookstores in Europe and America, and instantly rose to the ranks of a best-seller. Continuing the long-standing tradition of materialist scientism, Hawking ratchets the claims of a mathematical physics to hitherto unimaginable heights. Not content to assert, as other materialists have done, that whatever transpires in space and time can in principle be understood in physical terms, he argues that the very existence of the universe—"that there is something rather than nothing"—can likewise be explained on scientific grounds. Now, convinced as I am that this prestigious tractate by the "Einstein" of our day will all the more befuddle an already misinformed populace—that in fact these new "revelations from on high" are bound to cause immeasurable harm—I set about, persuaded by a friend, to write what I hoped would be the definitive response. Since the grounds for rebuttal had already been established in previous publications, beginning with *The Quantum Enigma*, and much of this material had in fact been summarized in *Science and Myth*, my objective could be realized in the concise format of an article. It is fitting, therefore, now that this task has been accomplished, to reprint that "Response to Stephen Hawking"—which meanwhile has appeared in the journal *Sophia*[1]—as a chapter in the last-mentioned book. The article has thus been added in this Second Edition: as Chapter 7, to be exact. The reader may observe for himself how every topic treated in the earlier chapters relates intimately to the ensuing "contra-Hawking" argument, and how the conclusions reached in chapters 2–5, in particular, vouch for its

1. Vol. 16, no. 2 (2011), pp. 5–48.

i

cogency. Even the material of chapter 6 bears vitally upon the Hawking issue: for in providing a glimpse, at least, of the traditional sciences, it renders visible the categorical limitations of contemporary physics at large by situating that science within a wider field.

It remains to thank Professor Seyyed Hossein Nasr, the aforementioned friend, for calling the Hawking book to my attention, and impressing upon me the urgency of refuting its disastrous claims: I am deeply grateful to him for this kindness.

Camarillo, February 22, 2012

INTRODUCTION

Science, according to the prevailing wisdom, constitutes the very antithesis of *myth*. As Albert Einstein has famously said, it deals with "*what is*"; in which case myth has to do, presumably, with "*what is not*." It turns out, however, that the matter is not quite so simple. In the first place, it happens that science does *not* refer purely and simply to "what is": even in the case of physics, its most accurate branch and foundational discipline, it refers, finally, not to Nature as such, but to the responses, on the part of Nature, to the strategies of the experimental physicist, which is something else entirely. Obviously, this was not understood in Newtonian times, and to this day is hardly ever acknowledged in our schools and universities; yet it is physics itself, in the form of quantum theory, that disqualifies our customary view of what it is that physics brings to light. Like it or not, physics deals—not simply with "*what is*"—but ultimately with what John Wheeler terms a "participatory universe." A disconnect, therefore, exists between what science itself affirms and what one generally takes to be the scientific world-view; in a word, that so-called scientific world-view turns out, finally, to be itself a *myth*.

We tend, however, to be equally confused regarding the nature and function of myth itself. We forget that, so far from dealing simply with "what is not," *authentic* myth "embodies the nearest approach to absolute truth that can be expressed in words," as Ananda Coomaraswamy points out. Yet in practice the two misconceptions—the over-valuation of science and the under-valuation of myth—go together, and count equally as a mark of enlightenment among the "well-informed." To complicate matters, science itself, as we have noted, begets myths of its own: a kind that would banish all others, and in so doing undermine, not only religion and morality, but indeed all culture in its higher modes. I say this without denigrating, in the slightest, the authentic achievements of

1

science: I deny neither the beauty and sublimity of its actual discoveries, nor the fact that the resultant technology, wisely utilized, can be of benefit to mankind. I speak rather of science in its present-day actuality as the prime determinant of culture: the oracle before which Western society at large has come to bow down in a kind of mindless adoration. How could it be otherwise, given that few, these days—a mere handful, it seems—distinguish with any degree of clarity between *science* and scientistic myth! Science has thus *de facto* turned into a kind of Trojan horse: we know not what we have let into our city. Seduced by the miracles of technology, we open ourselves to what we take to be scientific enlightenment, unaware of what it is that we imbibe; it is exactly as Christ foretold in his apocalyptic discourse to the disciples, when he spoke of "great signs and wonders" that might "deceive even the elect."

This monograph is concerned throughout with "science and myth." Its intended function, however, so far from being "merely academic," is eminently practical: the central and primary aim—of each chapter as well as of the whole—is to break the spell of *scientistic* myths, their stranglehold upon educated minds, and in so doing, to provide access once more to the perennial myths of mankind. These are the kind that *open* doors rather than bolt them shut, the kind that convey a sense of the *sacred*, which is finally none other than a sense of the Real. Contrary to what we have been taught to believe, the Real is not what we catch in our nets, but precisely what we do *not* catch: what ever eludes our mental grasp. It is, in a way, what ultimately "catches" us. And that is why it must be sought, figuratively speaking, "with folded hands," a gesture that betokens, not a "grasp," but the very opposite: a *submission*, namely, an unconditional openness, like that of a mirror wiped clean. But does this imply that there is nothing to be told of the Real: no doctrine at all? That, as I say, is where authentic myth comes into play: the kind that "embodies the nearest approach to absolute truth that can be expressed in words."

One thing more remains to be done by way of introduction: given that the book need not be read sequentially, it will be expedient to preview its content chapter by chapter. I shall be brief.

1: SCIENCE AND MYTH

This essay is designed to connect with Ananda Coomaraswamy's elucidations regarding the nature and function of authentic myth. It begins with the observation that science too is based upon "myths" (known as "paradigms"), and proceeds to enumerate the three which currently preside: the Newtonian, the Darwinian, and the Copernican. The first is simply the paradigm of "mechanism," which despite its spectacular success over several centuries was invalidated (as foundational) through the advent of quantum physics. The second is still dominant in biology, but hardly squares with the facts, and has moreover been disqualified by William Dembski's discovery of "intelligent design." The third—the so-called Copernican principle, which stipulates a constant average density of matter in space—still underpins contemporary astrophysics, but stands today on the verge of failure (due in part to seemingly insuperable difficulties in accounting for the formation of stars and galaxies). Now, what I wish to emphasize is not simply that these presiding paradigms stand on shaky ground and ought *de jure* to be replaced, but that they constitute in fact a species of myth, what I term "anti-myth." My major point is that these "myths of science"—each in its own distinctive way—militate against the perennial and indeed *sacred* wisdom of mankind.

2: MODERN SCIENCE AND GUÉNONIAN CRITIQUE

Here we reflect upon the Guénonian critique of modern science as it applies, in particular, to physics. Surprisingly enough, much of what the French metaphysician has to say in that regard proves to be plainly false, due to the fact that he conflates *true science* and *scientistic belief*. On the other hand, his conception of *quantity* as "the 'residue' of an existence emptied of everything that constituted its essence" turns out to be a master-stroke: the key, in fact, to the metaphysical understanding of modern physics, beginning with quantum theory. In light of considerations previously delineated in *The Quantum Enigma*, I present a philosophy of physics based upon the aforesaid Guénonian conception of "quantity."

3: SCIENCE AND EPISTEMIC CLOSURE

This chapter, too, deals with the philosophy of science—and of physics, especially—this time based upon the notion of "epistemic closure" introduced by Jean Borella, which may be defined as the elimination (from a concept) of all that is recalcitrant to expression in linguistic or "formal" terms. As Borella points out, this proves to be the defining condition of *scientific* as distinguished from *philosophic* thought. The latter is in fact characterized by what he terms "*l'ouverture à l'être*": the very opposite, namely, of epistemic closure. Following a brief introduction to Borella's thought, I show that these twin notions empower a philosophy of physics that rigorously accounts for the nexus between "science and myth," which in a way brings to completion my earlier studies in this field.

4: THE ENIGMA OF VISUAL PERCEPTION

The chapter presents a theory of visual perception propounded by the late James Gibson, based upon experimental findings accumulated over a period of several decades. It recounts how this hard-headed scientist was led, *on the basis of empirical facts*, to deconstruct the Cartesian dualism which underlies our scientistic world-view. What Gibson discovered is that *perception is not of a visual image* (be it retinal, cortical or mental)— as just about everyone had thought, at least since the days of Descartes— but that, on the contrary, we actually perceive what he calls "the environment" (in basically the ordinary sense). After delineating the main steps of Gibson's argument, I interpret his findings from a metaphysical point of view, and show, in particular, how his most provocative claims prove actually to be what I term "intellective features of visual perception": features, namely, that betoken "intellect" (*buddhi*) as distinguished from "mind" (*manas*).

5: NEURONS AND MIND

Here we begin by recalling the basic facts of contemporary neurophysiology, culminating in a description of the so-called primary visual system and an account of certain key experiments. This leads to a consideration of what is sometimes termed the "binding problem"—the question by what means "the computer is *read*": how the states of a million neurons give rise to a *single* object of perception or thought—and in particular, to the claim of Roger Penrose (the former mentor of Stephen Hawking) that the binding problem demands a theory of "quantum gravity." But whereas I find much in the thought of this great scientist to be of major interest (for instance, his demonstration, on the strength of Gödel's famous theorem, that computers cannot "do mathematics"), I contend that what the binding problem actually demands is, in essence, the Vedantic anthropology with its doctrine of the five *kośas*. After recalling the needful conceptions, I show how the relevant facts of neurophysiology can be integrated into the aforesaid doctrine.

6: *CAKRA* AND PLANET: O. M. HINZE'S DISCOVERY

In this chapter I report on a discovery by the German phenomenologist Oskar Marcel Hinze, which I regard as epochal. What stands at issue is a hitherto unsurmised isomorphism between macro- and microcosm, based on the Gestalt aspects of planetary astronomy and the *cakra* anatomy of man, as described in Kashmiri Tantrism. Given that each of the six principal *cakras* is associated with a symbolic *padma* or "lotus" as well as with corresponding letters of the Devanāgari alphabet (the number of which equals the number of "lotus petals"), and that each *cakra* is traditionally associated with a planet, Hinze set out to ascertain whether the "petal numbers" are manifested somehow in the phenomenology of the corresponding planetary orbits. He discovered not only that this is in fact the case, but that even the divisions of the corresponding Sanskrit letters into long and short vowels, sibilants, gutturals, palatals and cerebrals is faithfully reproduced on a planetary scale. We catch here a glimpse of *traditional* science in its unsurmised immensity, and of the gulf that separates these supposedly "primitive superstitions" from "science" as we conceive of it today. Yet what Hinze has to tell, so far from being "mythical," proves to be *scientific* by any count; and as I point out, it happens that his discovery actually invalidates our contemporary understanding of how the planetary system came to be. Yet this is not all: the slender treatise which brings to light the aforesaid isomorphism between "*cakra* and planet" culminates, surprisingly enough, in a ground-breaking essay on the doctrine of Parmenides. Suffice it to say that Hinze is to be ranked among that highly select group of authors—which includes Jean Biès and Peter Kingsley—who have begun to rediscover the true face of the Presocratics.

7: FROM PHYSICS TO SCIENCE FICTION: RESPONSE TO STEPHEN HAWKING

This chapter (which has been added in the Second Edition) presents an in-depth study of Hawking's supposedly scientific views concerning the nature and origin of the universe, as expounded in *The Grand Design*. Our analysis brings to light the hidden metaphysical and epistemological premises upon which the aforesaid claims are based, which prove in the end to be not only unfounded, but indeed untenable. It turns out, moreover, that virtually all the salient conceptions and findings presented in the first six chapters of this book enter naturally into these critical considerations, and in fact provide the grounds, both philosophical and scientific, for the ensuing refutation. Finally we argue that not only Hawking's thesis, but indeed the contemporary scientific world-view at large, proves to be ideology-driven, which is to say that in its cosmological claims it outstrips the evidence upon which it is reputedly based. Whether it be a question of Darwinist evolutionism or of the celebrated big bang, what ultimately stands at issue and tilts the scale is an *a priori* commitment to materialism, or more precisely, an unconditional denial of intelligent design.

8: METAPHYSICS AS "SEEING"

The final chapter is centered upon the idea of the "phenomenon" in the original sense of the Greek term, as "that which shows itself in itself." I point out that not only has this sense been lost, but that in the wake of Cartesian bifurcation the real is no longer conceived as the veritable phenomenon, but as something that stands *behind* "that which shows itself," something therefore which does *not* "show itself *in itself*." I then attempt to convey the gist of Edmund Husserl's so-called phenomenology (which I regard as perhaps the most outstanding contribution to philosophy in the twentieth century). This leads me to consider Goethe's approach to science and his critique of the Newtonian theory. I show that the former is in fact a phenomenology: that for Goethe true science was a matter of *Anschauung*, of "*seeing*" that which "shows itself in itself." I point out, moreover, that Goethe's deprecation of the Newtonian *Weltanschauung*—which hardly anyone, at the time, took seriously—has in fact been vindicated, even on a scientific plane, through the discovery of quantum theory. The following question arises now: having touched upon various levels of "seeing," ranging from the usual kind that actually "sees not" to the superior modes contemplated by phenomenologists, one is led to ask what is to be said of the ultimate or "absolute" seeing, which can be none other than a "seeing with the Eye of God." It is the response to this question that closes the chapter, and indeed the book; and here I base myself squarely on the teaching of Meister Eckhart, which goes to the very heart of the matter. What Eckhart gives us to understand comes down to this: *That which shows itself in itself—the veritable Phenomenon—proves to be, finally, none other than the Logos, the Word known to Christianity as the Son of God.*

1

SCIENCE AND MYTH

Third Ananda Coomaraswamy Memorial Lecture
The Sri Lanka Institute of Traditional Studies, Colombo, Sri Lanka

IT IS FITTING, in a Memorial Lecture honoring Ananda Coomaraswamy, to reflect upon the significance of "myth"; for indeed, it was the Sri Lankan savant who opened our eyes to what may be termed the *primacy of myth*. In one of his several masterpieces—a slender book entitled *Hinduism and Buddhism*—Coomaraswamy begins by recounting the mythical basis of the respective traditions before turning to their doctrinal formulations. He gives us to understand that myth exceeds doctrine, somewhat as a cause exceeds an effect or the original an artistic reproduction. It is not the function of doctrine to take us *out* of the founding myth: to "explain it away." On the contrary, its function is to bring us *into* the myth; for indeed, the pearl of truth resides in myth as in a sanctuary. Authentic doctrine can take us to the threshold of that sanctuary; but like Moses before the Promised Land, it cannot enter there.[1]

Not all doctrine, however, is sacred, and it turns out that atheists and iconoclasts have myths of their own. Not only the wise, but fools also live ultimately by myth; it is only that the respective myths are by no means the same.

My first objective will be to exhibit the mythical basis of modern

---

1. Theologians may contest the primacy of myth in the case of the so-called monotheistic religions, on the grounds that in these traditions historical fact has replaced myth. Yet nothing prevents historical fact from being also a myth. The "primacy of myth" attains actually its highest reading in the founding fact of Christianity, when "the Word became flesh, and dwelt among us . . . ." (John 1:14)

science. In particular, I shall discuss three major scientific myths (generally referred to as "paradigms"): the Newtonian, the Darwinian, and the Copernican. My second objective will be to contrast the myths of Science with the myths of Tradition. I will voice the conviction that this discernment is of great moment, that indeed it vitally affects our destiny, here and hereafter.

I

There was a time when science was thought to be simply the discovery of fact. It is simply a fact, one thought, that the Earth rotates around the Sun, that force equals mass times acceleration, or that an electron and a positron interact to produce a photon. It was as if facts "grew upon trees" and needed only to be "plucked" by the scientist. In the course of the twentieth century, however, it was discovered that this customary assumption is not actually tenable. It turns out that facts and theory cannot be ultimately separated, that "facts are theory-laden," as the postmodernists say. The old idea that the scientist first gathers facts, and then constructs theories to explain these facts, proves to be oversimplified. Behind every science there stands perforce a paradigm—a "myth" one can say— which determines what is and what is not recognized as a fact. When Joseph Priestley, in 1774, heated red oxide of mercury and collected a gas known today as "oxygen," did he actually discover oxygen? So far as Priestley himself was concerned, he had found "dephlogisticated air"! To discover oxygen, something more is needed, besides a vial of gas: a theory, namely, in terms of which that gas can be interpreted or identified. Not until, a few years later, Lavoisier had constructed such a theory did oxygen (or the existence of oxygen, if you prefer) become an established scientific fact.

Just as, in the words of Wittgenstein, thought never gets "outside language," so too a science never gets "outside" its own paradigm. It is true that paradigms are sometimes discarded and replaced; this happens, according to the historian and philosopher Thomas Kuhn, in the wake of crisis, when the presiding paradigm can no longer accommodate all the findings to which, in a sense, it has led. But though a science may indeed outgrow a particular paradigm, it never outgrows its need for such: in a word, the "mythical element" in science cannot be exorcised. And indeed, the moment science denies its "mythical" basis it turns illusory, and thus becomes "mythical" in the pejorative sense of that term.

The first of the three "presiding paradigms" I have singled out is the Newtonian, which defines the notion of a mechanical world or clockwork universe. What exists, supposedly, is "bare matter," the parts of which interact through forces of attraction or repulsion so that the movement of the whole is determined by the disposition of the parts. To be sure, the concept of "bare matter"—the Cartesian notion of *res extensa*—is philosophically problematic, and hinges moreover upon the Cartesian postulate of "bifurcation": the idea, namely, that all qualities (such as color) are subjective, and that, consequently, the external object is not in fact perceived. It will be recalled that Descartes himself felt obliged to prove—by means of a now famous argument of questionable cogency—that even though the external world proves thus to be imperceptible, it nonetheless exists. One may further recall that twentieth century philosophy, on the whole, has veered away from the Cartesian position, and that "bare matter," in particular, has been downgraded to the status of an abstraction. To take *res extensae* for the real—as scientists are wont to do—is to commit what Whitehead terms "the fallacy of misplaced concreteness": it is to mistake a concept for a reality. What presently concerns us, however, is not the philosophic validity of the Newtonian paradigm, but its scientific efficacy, which is quite another matter. Though the Newtonian worldview may indeed be spurious—a "myth" in the pejorative sense of this equivocal term— history confirms that it has nonetheless functioned brilliantly in its capacity as a scientific paradigm. It appears that error too has its use! One sees in retrospect that science of the contemporary kind could never have "lifted off the ground" without the benefit of a worldview that is drastically oversimplified, to the point of being incurably fallacious.

Despite its philosophical invalidity, the success of the Newtonian paradigm has been spectacular. From the publication of Newton's *Principia*, in the year 1687, to the beginning of the twentieth century, it was universally regarded, not simply as a successful paradigm, but indeed as the master-key to the secrets of Nature, from the motion of stars and planets to the functioning of her minutest parts. I will not recount the triumphs of Newtonian physics which seemingly justified this grand expectation; the list is long and singularly impressive. Suffice it to say that by the end of the nineteenth century the Newtonian scheme had extended its sway beyond the bounds of mechanics, as commonly understood, to include electromagnetism, which, as it turns out, cannot be pictured in grossly mechanical terms. Yet even here, in this "aetherial" domain, the notion of a whole rigorously reducible to its infinitesimal parts has proved once again to be key: the justly famous Maxwell field equations testify to this fact. What is more, even the revolutionary proposals of Albert Einstein, which did break with some of the basic Newtonian conceptions, have left the foundational paradigm intact: here too, in this sophisticated post-Newtonian physics, we are left with a physical universe which can in principle be described with perfect accuracy in terms of a system of differential equations. In a vastly extended sense, the Einsteinian universe is still *mechanical*, which is to say that it conforms precisely to what we have termed the Newtonian paradigm.

What finally dethroned that seemingly invincible paradigm was the advent of quantum mechanics, which proves not to be a mechanics at all: the whole, it now turns out, is not in fact reducible to its parts, be they finite or infinitesimal. At the same time—and in consequence of this irreducibility—the new so-called mechanics turns out not to be fully deterministic. It is no longer possible, in general, to predict the exact value of an observable; instead, the problematic notion of "probability" has now entered the picture in a fundamental and irreplaceable way. This is what Albert Einstein—the greatest and loftiest among the advocates of mechanism—could not bring himself to accept; the idea that "God plays dice," as he put it, was simply abhorrent to him. Thus, till the end of his life, he staunchly refused to accept quantum theory as something more than an approximation. Yet everything we know today does point to the fact that it is actually relativistic mechanics that proves to be "merely approximate," whereas quantum theory appears to be fundamental. This is not to say that the picture may not change; but whatever the future may bring, it is safe to surmise that a return to mechanism is not in the cards.

We turn now to the Darwinian paradigm, which proves to be, in a sense, the opposite of the Newtonian: for it happens that Darwin's idea has been an unmitigated failure from the start. I contend, in fact, that Darwinism is not in truth a scientific theory, but is simply an ideological postulate masquerading in scientific garb. To be sure, given the imposing prestige and unending encomiums lavished upon this doctrine by the academic and media establishments alike, these claims are of course surprising; but let us take a look at the facts of the case.

Darwin claims that existing species are derived from one or more primitive ancestors through chains of descent extending over millions of years. Never mind, for the moment, by what means the stipulated transformation from primitive to differentiated organisms may have come about; whatever the means, it is clear that Darwin conceived of this evolution as a gradual process involving countless intermediary forms, many if not most of which should by right appear in the fossil record. Yet apart from a handful of doubtful specimens, intermediary types are nowhere to be found. This fact is now generally admitted even by scientists who believe in some kind of evolution. Steven Jay Gould, for instance, one of the foremost authorities, has felt compelled to abandon orthodox Darwinism for precisely this reason. "Most species exhibit no directional change during their tenure on earth," he writes. "They appear in the fossil record looking pretty much the same as when they disappear; morphological change is usually limited and directionless."[2] One would think that this alone suffices to disqualify the transformist hypothesis; but to disciples of the British naturalist, it merely implies that evolution must take place at such speed, or under such conditions, that the intermediary forms disappear without leaving a trace. As

2. Quoted by Phillip Johnson in *Darwin on Trial* (Downers Grove, Illinois: Intervarsity Press, 1993), p. 50.

Phillip Johnson, the Berkeley law professor and author of *Darwin on Trial*, has observed: "Darwinism apparently passed the fossil test, but only because it was not allowed to fail."

Darwin's great idea, it will be recalled, is that Nature produces small random mutations, which are passed on to the genetic line in accordance with the phrase "survival of the fittest." It has been pointed out that this famous dictum, which supposedly constitutes the key to the riddle of evolution, is in fact a tautology, much as if to say "the rich have plenty of money"; this, in any case, is what the philosopher Karl Popper meant when he declared Darwin's theory to be "unfalsifiable," and therefore void of scientific content. Falsifiable or not, however, Darwin's doctrine does stake a claim. So far from being true by definition, it constitutes in fact one of the most astronomically improbable conjectures ever conceived by the mind of man. Take the case of an eye, for example: Darwin is telling us that this structure of almost unimaginable complexity came about through a series of minute accidental mutations. Leaving aside the circumstance that a rudimentary eye which can not yet see is of no use whatever in the struggle for survival, calculations carried out by the mathematician D.S. Ulam show that the requisite number of mutations turns out to be of a magnitude so immense that, even within a time frame measuring billions of years, the likelihood of such an occurrence is vanishingly small. But this too does not pose a problem for the committed Darwinist; as Ernest Mayr has said by way of response: "Somehow or other by adjusting these figures we will come out all right. We are comforted by the fact that evolution has occurred."[3] And this is indeed the crucial point: for the dyed-in-the-wool Darwinist, evolution as Darwin conceived of it is itself the most indubitable fact.

There are those who claim that recent advances in molecular biology have at last supplied hard evidence in support of evolution. Now, it is true that the findings in question permit us to quantify the "molecular distance," so to speak, between genomes, and thus between species. Moreover, given the fact that mutations occur at a more or less constant rate, it is possible to estimate the time

3. Ibid., p. 38.

required to effect a given genetic alteration, as measured by the aforesaid distance. If, therefore, two species have descended from a common ancestor, one can now estimate how long ago the stipulated separation must have occurred. On this basis one speaks nowadays of a so-called molecular clock, which supposedly measures the rate at which evolution takes place. However, in the euphoria generated by this discovery, one forgets that not even a "molecular clock" can measure the rate of evolution unless evolution has indeed occurred. But this hypothesis remains today as unconfirmed as it has been from the start. What is more, it turns out that the findings of molecular biology are not in fact propitious to the evolutionist cause: the very precision with which molecular structures and processes can now be understood spells trouble for the Darwinist. This is what Michael Behe, the now famous molecular biologist, has demonstrated so forcefully in *Darwin's Black Box*, a book which has decisively affected the debate.

To cite at least one example of amazing facts adduced by Behe, I will mention the so-called bacterial flagellum,[4] a kind of paddle used to propel the bacterium through its liquid ambience, driven by a molecular rotary engine, which is powered by an acid. The structure is exceedingly complex, and involves about two hundred and forty different kinds of proteins, which need all to be in place if the engine is to function and the flagellum is to do its job. We have here an example, on a molecular scale, of what Behe terms irreducible complexity. "By *irreducibly complex*," he explains, "I mean a single system composed of several well-matched, interacting parts that contribute to the basic function, wherein the removal of any one of the parts causes the system to effectively cease function."[5] The notion proves to be crucial: it is not in fact possible to account for the genesis of irreducibly complex structures in Darwinist terms. This can now be demonstrated by means of design theory, a mathematical discipline which allows us to conclude that no process compounded of "chance" and "necessity" can give rise to irreducible complexity, or to something still more general termed complex

---

4. *Darwin's Black Box* (New York: The Free Press, 1996), pp. 70–73.
5. Ibid., p.39.

specified information.[6] The new mathematical theory, in conjunction with the sharp data of molecular biology, provides at last a rigorous refutation of Darwin's hypothesis. Of course, whether even this will convince the die-hard Darwinist remains to be seen. Meanwhile, after more than a decade of debate and controversy regarding "intelligent design," it seems that the Darwinist establishment—aided by the media—has succeeded brilliantly in confusing the issue literally beyond recognition: victory by obfuscation, one might say.

Our third paradigm pertains to contemporary cosmology. It happens that field equations plus astronomical data do not suffice to determine the global structure of the physical universe: an infinite number of "possible worlds" remain. One therefore requires an additional hypothesis. Following Einstein's lead, scientists have generally opted for a condition of spatial uniformity in the distribution of matter: one defines an average density of matter, and assumes this to be constant throughout space. Thus, on a sufficiently large scale, the cosmos is thought to resemble a gas, in which the individual molecules can be replaced by a density of so many grams per cubic meter. It was Hermann Bondi who first referred to this assumption as the Copernican principle, and not without reason; for even though Copernicus himself knew nothing about a supposedly constant density of stellar matter, the principle in question constitutes in a way the ultimate repudiation of geocentrism, and thus consummates what has been termed the Copernican revolution. Henceforth space in the large is assumed to be void of structure or design, and subject only to local fluctuations from an average density, much like the molecular fluctuations in a gas, which remain imperceptible on a macroscopic scale. I would like

6. The mathematics of design theory has been expounded in William A. Dembski's *The Design Inference* (Cambridge University Press, 1998). For its implications regarding Darwinism, see Dembski's *Intelligent Design* (Downers Grove, Illinois: Intervarsity Press, 1999).

however to impress upon you that this is not a positive finding of astrophysics or a proven fact, but simply an assumption: to be precise, it is the postulate or hypothesis which underlies our contemporary scientific cosmology.

It was Einstein who first proposed such a "universe" by postulating an average density of matter which is constant, not only in space, but also in time. He discovered, however, that his field equations admit no such solution unless one adds an additional term involving the so-called cosmological constant. Thus, to prevent his static universe from collapsing under the influence of gravity, Einstein did add the term in question. Before long, however, a Russian mathematician, named Alexander Friedmann, succeeded in showing that solutions to Einstein's field equations can be obtained without this *ad hoc* constant, simply by letting the stipulated density of matter vary with time. What Friedmann had discovered mathematically was an expanding universe, a cosmos of the big bang variety. Not long thereafter Edwin Hubble, an American astronomer, arrived at substantially the same conclusion on the basis of astronomical findings[7]; and eventually Einstein himself acceded to the notion of a time-dependent universe. Discarding the cosmological constant—"the biggest mistake of my life" he called it—he joined his colleagues in accepting the scenario of an expanding universe, said to have emerged out of an initial singularity some fifteen billion years ago.

Before too long, however, big bang cosmology ran into difficulties, which have since led to a number of modifications in an ongoing effort to accommodate the mathematics to the empirical data of astronomy. Nonetheless, all is not well, and those who claim otherwise "overlook observational facts that have been piling up for 25 years and have now become overwhelming," as Halton Arp pointed

---

7. Hubble's conclusion is based on the phenomenon of "red-shift" in stellar spectra, which he interprets as a Doppler effect. That assumption, however, is not only unfounded, but has in fact come under attack in recent years, due to an abundance of adverse empirical evidence. See Halton Arp, *Seeing Red* (Montreal: Apeiron, 1998). On the scientific basis of big bang cosmology, I refer also to my treatise *The Wisdom of Ancient Cosmology* (Oakton, VA: Foundation for Traditional Studies, 2003), chap. 7.

out in 1991. For example, astronomers claim to have spotted galaxies separated by close to a billion light-years. Now, given the low relative velocities observed between galaxies, it would take about 200 billion years to arrive at such a separation from an initially uniform state: a good ten times longer than the estimated age of the universe. Or, to cite another fundamental difficulty: there seems not to be nearly enough matter in the universe to generate gravitational fields strong enough to account for the formation and persistence of galaxies. Such incongruities, however, are generally taken in stride by the experts. As Thomas Kuhn points out, the primary concern of "normal science" is to preserve the paradigm, to protect it, so to speak, against hostile data. What does one do, for instance, if there is not enough matter in the universe to account for galaxies? One strategy is to introduce something called "dark matter," which supposedly does not interact with electromagnetic fields, and is consequently invisible. Its only measurable property is gravitation, and its only discernible effect is to bring the gravitational field up to the levels demanded by the big bang scenario. Never mind that not a single particle of dark matter has ever been detected: for advocates of big bang theory, it seems, the existence of galaxies is proof enough. According to some estimates, proposed by respected members of the astrophysical community, about 99% of all matter in the universe is dark. What is more, one postulates two kinds of dark matter: so-called "hot" and "cold," with very different properties, in a mix of ⅓ hot and ⅔ cold as the required blend!

Other parameters of questionable authenticity have likewise been enlisted in the defense of big bang theory. The cosmological constant, for example, turns out to be of use after all: it has thus been claimed that the resurrected constant accounts for about 80% of the estimated energy density. Strangely enough, the parameter postulated to explain why Einstein's static universe does not collapse serves now to explicate why galaxies don't fly apart.

Yet, despite an abundance of theoretical options for coping with troublesome data, it appears that big bang cosmology is approaching a state of crisis. A growing number of scientists concur with Halton Arp to the effect that adverse facts have been piling up, and that a point has now been reached, beyond which defense of the paradigm is no longer compatible with sound scientific practice. It remains to be seen whether the Copernican paradigm will weather the storm.

II

The tenacity and fervor with which the presiding paradigms of science are defended even in the face of plainly hostile data suggest that, here too, an element of ideology may be at play. Science is not in reality the purely rational and "disinterested" enterprise it pretends to be; it is after all the work, not of computers, but of men. There is reason to believe that the paradigms of science are more in fact than cold, sober conjectures, mere hypotheses to be discarded in the face of contrary evidence. It appears that the top paradigms, at least, are weightier by far than that. In addition to their formal or "operational" connotation, one finds that these paradigms carry a wider sense, a "cultural" meaning, one can say; and it is mainly this broader connotation, which actually eludes scientific definition, that mainly communicates itself to the public at large, which in fact is incapable of comprehending its strictly "scientific" use.

Now, it is this circumstance that in a way justifies our claim that science entails an element of "myth." I say "in a way," because it happens that traditional or authentic myth is something far greater, something that categorically exceeds the "mythical" dimension of scientific paradigms. Let us say, then, that there are different *kinds* of myth, ranging all the way from the sacred to the profane, from the sublime to the trivial or absurd. We need, moreover, to understand that man does not live by "fact," or by "fact" alone, but preeminently by "myth": this is indeed, culturally speaking, his daily "bread." What, above all, differentiates one man from another—again, from a "cultural" point of view—is the presiding myth that directs, motivates, and informs his life. I contend that the stature and dignity of a person depend primarily on the myth he has made his own; in a way we become what we believe. And I would add: no more telling reason has ever been proposed for treading cautiously!

To comprehend the nature and function of "myth," we need, first of all, to get over the idea that myth has to do with what is imaginary or unreal, a notion which came into vogue in the course of what historians call the Enlightenment, when men thought that science had at last delivered us from the childish dreams of a primitive age. In this optic, myth was perceived simply as the antithesis of fact: at most a pleasurable or consoling fiction. One might go so far as to admit that such fictions may be indispensable: that our life would be intolerably drab and void of hope without some kind of mythical embellishment; but when it comes to the question of truth, it is to Science that we must look.

Such then was the prevailing view of myth during the age of modernism; but that phase, as one knows, is presently nearing its end, both philosophically and culturally. The new outlook, generally termed postmodernist, breaks with the old: the deconstructionist zeal, which in days gone by was directed mainly against established religious, cultural, and political norms—against everything, one could say, that smacked of tradition—has now been turned against the scientific enlightenment as well. There is logic in this, and a certain justice too; but yet it needs to be understood that the effects of the Enlightenment or modernity upon our *Weltanschauung*—and in particular, on our ability to perceive what science is actually about—have not been thereby canceled or ameliorated. Readers of Ananda Coomaraswamy will comprehend very clearly how much we have lost: that despite the material advantages of modern life, we have become woefully impoverished. In fact, we have arrived at the point of losing what is truly "the one thing needful." Cut off—as never before—from the source of our being, we have all but forgotten that life has meaning: a goal and a possibility which is *not* ephemeral; but needless to say, neither modern science nor its postmodernist critics can enlighten us in that regard. *For this one requires authentic myth*: the kind that belongs inextricably to sacred tradition as the paramount expression of its truth. Such myth, says Ananda Coomaraswamy, "embodies the nearest approach to absolute truth that can be stated in words"[8]: a far cry indeed from the prevailing conception of myth as "the fictitious"!

Myth alone, however—no matter how exalted it may be—will

---

8. *Hinduism and Buddhism* (Westport, CT: Greenwood Press, 1971), p. 33.

not save, liberate, or enlighten us. Traditionally speaking, the illuminating myth must be received under appropriate auspices, which include conditions upon the recipient or disciple, the chief of which is *śraddha*, faith: there can be no spirituality, no true enlightenment, without faith.

Now, it is at this point, I say, that modern science touches upon the spiritual domain: it enters the picture, I contend, not as an ally of true religion, but perforce as an impediment to faith, and therefore as a spoiler, an antagonist. It is a case of opposing myths, of mythologies that clash: or better said, *of myth and anti-myth*.

Let us try to understand this clearly. We must not be put off by the simplistic look of traditional myth, its typically crude literal sense, remembering that such myth speaks, not to the analytic mind, but to the intuitive intellect, sometimes termed "the eye of the heart," a faculty which, alas, modern civilization has been at pains to stifle. Now, it is precisely on this level of understanding— the level of the authentic Intellect—that myth does in fact constitute "the nearest approach to absolute truth." What we have termed the "myths" of science—namely, its paradigms, be they true or false—on the other hand, deliver such content as they have primarily to the rational mind; there is no mystery here, no reference to higher realms of truth. Quite to the contrary: these so-called myths offer a substitute, a "quasi-truth" here below, a kind of idol of the mind, which impedes our spiritual vision. As a tool of science—as a paradigm in the strict sense—they have of course a legitimate use: think, for instance, of the now discredited Newtonian paradigm. The trouble with paradigms, however, is that they tend to become absolutized, that is to say, dissociated from the scientific process; and this is where the idolatry sets in. One transitions surreptitiously from the hypothetical to the certain, from the relative to the absolute, and thus from a science to a metaphysics. But not to an authentic metaphysics! True to its origin, that "relative rendered absolute" remains unfounded and illegitimate, a pseudo-metaphysics one can say. It needs to be understood that a paradigm of science *absolutized* turns forthwith into an anti-myth.

I realize that in taking this stand I am offending against the "political correctness" of our day. We are told that the proverbial conflict between science and religion is based upon antiquated ideas. It has been said that in the age to come the two disciplines will be seen as complementary aspects of a single enterprise, each contributing to the good of man within its own appropriate sphere. All truth, we are assured, is ultimately consonant. But in midst of this idyllic harmony, it is always religion in its traditional modes that is obliged, by the presiding authorities, to conform itself to the putative truths of science by "demythologizing" its beliefs. One forgets that science too has its mythology, and that the putative truths at issue are not, strictly speaking, scientific or "operational," but pertain to its mythical side. The most obvious example is the Darwinist account of man's origin, which in fact has no "operational" content at all, and is consequently purely mythical. The problem, however, is that this "myth of science" is flatly opposed to every cosmogonic myth of sacred provenance, from the Vedas to the Book of Genesis. It appears that the "demythologizers" of religion do have a point! My complaint is that they are demythologizing the wrong thing: their intent is to jettison the sacred for the profane. In the name of this or that pseudo-myth, these blind guides have cast out "the nearest approach to absolute truth that can be stated in words." The new irenic approach to the old problem proves thus to be deceptive: the kiss of science, I say, is the death of religion.[9]

The conflict of which I speak calls to mind the implacable antagonism between the *Devas* and the *Asuras* ("gods" and "demons," good angels and bad) as depicted in Hindu lore; and I would add that the Darwinist doctrine, in particular, may be classified as distinctly *asuric* in content, and perhaps in its provenance as well. The Darwinist "myth" is in fact expressive of the *asuric* credo as formulated in the *Bhagavad Gita*:

---

9. Of authentic religion, that is. Drop that qualification, and my statement becomes patently false. We now find ourselves in the so-called New Age, the era of pseudo-religions, many of which (if not all) are indeed the offspring of the aforesaid unholy union. For a case-study pertaining to Christianity, I refer to my monograph on the teachings of Teilhard de Chardin. See *Teilhardism and the New Religion* (Rockford, IL: TAN Books, 1988; revised edition, *Theistic Evolution: The Teilhardian Heresy*, forthcoming from Angelico Press, 2012).

They say: "The world is void of truth, without a moral basis, and without a God. It is brought about by the union of male and female, and lust alone is its cause: what else?"[10]

Would it be too much to say, from a Christian point of vantage, that Darwinism stands on the side of Antichrist, the Father of Lies and primeval Antagonist of man's salvation?[11] We are dealing, in any case, not simply with beliefs or speculations of erring mortals, but with something far greater and incomparably more perilous; in the words of St. Paul: "We wrestle not against flesh and blood, but against principalities, against powers, against the rulers of the darkness of the world, against spiritual wickedness in high places."[12] It follows that the individual outside the pale of sacred tradition stands little chance of emerging from this contest unscathed. No matter how erudite or brilliant, even, we may be, our position is then at best precarious: far more hazardous, in fact, than we can normally imagine. It is no small thing to fall prey to *asuric* myth!

The case of Darwinism is admittedly exceptional; as we have had occasion to observe, the Darwinian paradigm stands out even from a scientific point of view both by its failure to accord with the observable facts as well as by the astronomical improbability of its claims. But what about the other paradigms of contemporary science: are these likewise opposed to the traditional worldview? There are of course a great number of paradigms in scientific use at the present time; the structure of science in our day is exceedingly

---

10. Chapter 16, verse 8. Having thus formulated the *asuric* credo, the *Gita* proceeds to describe the men who have made that creed their own: "Holding such a view, these lost souls of little understanding and fierce deeds rise up as the enemies of the world for its destruction." One cannot but think of the technocrats who will be "running the world" under the New World Order!

11. This view has been forcefully propounded by the late Orthodox hieromonk Seraphim Rose. See his masterful monograph, *Genesis, Creation and Early Man* (Platina, CA: St. Herman of Alaska Brotherhood, 2000).

12. Eph. 6:12.

complex, and literally entails "paradigms within paradigms." However, it is the top-level paradigms that matter most both from a philosophical and a cultural point of view; it is these that mainly define what we deem to be the scientific worldview. And that *Weltanschauung* is in fact characterized by the three paradigms we have singled out: the Newtonian, the Darwinian, and the Copernican. It is these, I say, that encapsulate our scientistic understanding of the physical, the biological, and the stellar world, respectively. To be precise: it is the Darwinian paradigm that enables us to extend the Newtonian into the biosphere—not legitimately, to be sure, but in some more or less imaginative fashion—and it is the Copernican that enables us to do likewise with reference to the stellar universe. It is thus by means of the Darwinian and the Copernican paradigms in conjunction that physics claims sway over all that is thought to exist in space and time.

This brings us to my final contention, namely, that all three of these top paradigms are in fact irreconcilably opposed to the traditional worldview. Having already identified Darwinism as an inherently *asuric* myth, it remains now to consider the Newtonian and Copernican claims. I must of course be brief; but I shall try in either case to touch upon the crux of the matter.

It is easy to see that there could be no such thing as *spiritual* life in a mechanical universe, because in such a universe there could in fact be no life at all: not even an ameba could exist in a Newtonian world! And why not? For the simple reason that no living organism is reducible to the sum of its parts. This fact has been well understood by philosophers at least since the time of Aristotle, and is being rediscovered and reemphasized today by some leading biologists. Traditional cosmologies, on the other hand, refer—not to some philosophical abstraction or scientific "model"—but to the authentic cosmos, the world in which we find ourselves, which not only serves as a habitat to plants and animals, but houses artists and poets, mystics and saints as well. Now, so far from constituting a mechanical system, the authentic universe constitutes in truth a theophany: a manifestation of what the Vedas term *nāma*, Plato terms Ideas, and St. Paul "the invisible things of God"—not forgetting that to the pure in heart it mirrors "even His eternal power and

Godhead."[13] There could in fact be no greater disparity between the cosmos, as traditionally perceived, and a Newtonian world: the two, it turns out, are not merely incompatible, but indeed antithetical. Thus, whereas the former exceeds what we are able to grasp by virtue of its inexhaustible fullness, the latter eludes our grasp on account of its emptiness, an indigence which literally defies imagination: for it must not be forgotten that the Newtonian world is perforce bereft of all qualities, beginning with color, and is consequently imperceptible. It constitutes a world (if such it may still be called) which can be neither seen nor imagined, and which consequently does not in truth answer to a "worldview" at all: whatever the scientistically indoctrinated public imagines the universe to be deviates *ipso facto* from the *scientific* contention. As is in fact the case with all doctrine of an *asuric* kind, the mechanistic worldview constitutes, finally, *a lie.*

The insufficiency of the Copernican paradigm is perhaps harder to discern, since it pertains to things remote in space and time, and thus remote from the familiar world. One needs however to recall that the sun, moon, and stars play a major role in the traditional worldview; as we read in a famous psalm of David: "The heavens declare the glory of God; and the firmament sheweth his handywork."[14] According to the Copernican principle, however, the cosmos at large exhibits no global structure, no hierarchic architecture, no trace of exemplarism or design: only matter randomly distributed, like so many atoms in a gas. Thus, while the Darwinian paradigm denies God as the Creator of life, the Copernican denies Him as the Architect of the universe. The hypothesis of constant average mass density throughout space may be a useful device for obtaining solutions to the field equations, but is hardly compatible with the perennial wisdom of mankind.

Fortunately, however, science is self-corrective to a degree, which is to say that in due time faulty paradigms are normally replaced. The

---

13. Rom. 1:20.
14. Ps. 19:1.

Newtonian has already been superseded, and both the Darwinian and the Copernican are now under attack. It may be true, as Thomas Kuhn maintains, that failed paradigms are invariably retained until a new one has been approved by the scientific community; but in the end this does apparently take place—so long, at least, as the scientific establishment retains a modicum of integrity. Science, as we know, constitutes an ongoing process, and even its most prestigious paradigms are by no means sacrosanct.

The only things sacrosanct, in fact, are the core elements of sacred tradition. It is the distinctive characteristic of sacred tradition to have a more-than-human, more-than-merely-historical origin, implying that authentic tradition, in all its essential manifestations—from doctrine and ritual to moral codes—partakes somewhat of eternity. We may accept or reject sacred tradition: that is our inalienable option; but let us understand from the start that outside of the Sacred there can be no certainty, no absolute and abiding truth.

2

MODERN SCIENCE AND GUÉNONIAN CRITIQUE

*To dismiss partial modes of knowledge simply because
they are what they are is just as grave a fallacy as to
mistake the partial for something total and all-embracing.*
Gai Eaton

READING René Guénon's discourse on modern science more than half a century after it was written, one is struck not only by the depth of its penetration, but also, to a lesser degree, by its glaring insufficiencies. My intention in the present chapter is to examine the Guénonian critique in reference to contemporary physics, especially quantum theory, its foundational discipline and most accurate branch.

I will begin by recalling René Guénon's diagnosis of the present age as "the reign of quantity." In this basic recognition he has not only characterized the prevailing scientific purview, but has, at the same time, interpreted its "reign" in light of a traditional *metaphysical* understanding of history. "Reduction to the quantitative," Guénon maintains, "is strictly in conformity with the conditions of the cyclic phase at which humanity has now arrived."[1] In keeping with Hindu doctrine, he envisages a descent "which proceeds continuously and with ever-increasing speed from the beginning of the *Manvantara*," and which, metaphysically speaking, constitutes "but a gradual movement away from the principle which is necessarily

1. *The Reign of Quantity and the Signs of the Times* (San Rafael CA: Sophia Perennis, 2004), p. 3.

25

inherent in any process of manifestation." In Christian terms, this corresponds to the Fall, conceived now as an ongoing process. "In our world," Guénon goes on to explain, "by reason of the special conditions of existence to which it is subject, the lowest point takes on the aspect of pure quantity, deprived of every qualitative distinction." The rise of modern science and its progressive domination of our culture comes thus to be perceived in metaphysical terms, and therefore from the deepest and most inclusive point of view.

With a kind of mathematical exactitude, Guénon delineates the multiple manifestations of that "descent to the lowest point" and the successive stages through which humanity is destined to pass. He alludes to the sheer blindness that constitutes both a precondition and a manifestation of the ongoing descent. "If our contemporaries as a whole," Guénon avers, "could see what it is that is guiding them and where they are really going, the modern world would at once cease to exist as such." It is no wonder that this world, despite its protestations of openness to dissenting views of every description, is in fact all but closed to the voices of tradition. "It is impossible," Guénon tells us, "that these things should be understood by men in general, but only by the small number of those who are destined to prepare in one way or another the germs of the future cycle." According to this view, the rise to dominance of the physical sciences both manifests and imposes "the reign of quantity."

However, along with such major recognitions—which I find unprecedented and indeed definitive—there are aspects of the Guénonian doctrine that strike me as less felicitous. I charge that these questionable tenets are not only gratuitous—that is to say, uncalled for on the basis of Guénon's central contentions—but demonstrably false. What primarily invalidates the Guénonian critique, as it pertains to physics in particular, is the failure to recognize that in the midst of what is admittedly a "scientific mythology," there stands nonetheless a "hard science," a science capable of an actual knowing, "partial" though it be. As I have argued repeatedly, the one thing most needful for a just appraisal of modern science is the distinction between "scientific knowledge" and "scientistic belief," that is to say, between *science*, properly so called, and *scientism*. Yet it appears that nowhere does Guénon draw that crucial distinction, apparently for the simple reason that he does not credit contemporary science with any *bona fide* knowledge at all. Admittedly, science and scientism are invariably joined in practice, and prove indeed to be *de facto* inseparable; whosoever has moved in scientific circles will have no doubt on that score. It can even be argued that scientistic belief plays a vital role in the process of scientific discovery, that in fact it constitutes a pivotal element in the scientific quest. Yet, even so, I maintain that the two faces of the coin are as different as night and day, and need to be sharply distinguished. As regards physics, in particular, I contend that there exists a body of positive findings that is logically independent of scientistic belief, and qualifies as a "a partial mode of knowledge," to put it in Gai Eaton's words. It is this *bona fide* knowledge, obviously, that powers the ongoing technological revolution, and thereby bestows upon science, in the eyes of the public, its immense authority and prestige. The fact that the public at large, and to a considerable extent the scientific community itself, confounds that knowledge with scientistic belief is another matter, about which we shall have more to say in the sequel. For the moment, I wish only to make the point that there is, in the modern world, such a thing as "hard" science, a discipline capable of positive findings, again *"partial" though these findings be.*

Yet as I have said, this is something Guénon seems never to admit. He distinguishes, as one must, between "the domain of a mere observation of facts" and the formation of hypotheses, but seems to regard the latter as void of cognitive value, that is to say, as void of truth. "The ever-growing rapidity," he writes, "with which such hypotheses are abandoned these days and replaced by others is well known, and those continual changes are enough to make all too obvious the lack of solidity of the hypotheses and the impossibility of recognizing in them any value so far as real knowledge is concerned."[2] Guénon's conclusion, however, is far from obvious, and proves in fact to be untenable. Take the case of physics: to be sure, the history of that science, from the time of Galileo to the present day, displays a succession of hypotheses; yet to perceive these alterations as a kind of wanton stabbing in the dark, a process which

2. Ibid., pp. 120–121.

achieves no lasting result and carries no value "so far as real knowledge is concerned," is surely to miss the point. What Guénon ignores is the fact that physics *evolves*, and that hypotheses are not simply "jettisoned," but are generalized and supplemented in light of new discoveries. Newtonian physics, in particular, was not just abandoned as an erroneous theory, but remains in constant use to this day, and is in fact rigorously implied by both Einsteinian relativity and quantum mechanics in the limit as c tends to infinity and h tends to zero, respectively. There is no question here of a haphazard "jumping to conclusions" as Guénon seems to suggest; it is a matter, rather, of physics becoming progressively more refined, more accurate, and more powerful in its applications. It is plain, moreover, that this evolution of physical theory is faithfully reflected in the concomitant development of technology; one needs but to compare steam engines to jet planes and spacecraft to behold, as in an icon, the increase in physical knowledge that has taken place. Quite frankly, one is amazed at the superficiality of Guénon's analysis when it comes to the *positive* side of contemporary science, and thus to contemporary science as distinguished from the accompanying scientism; one can only surmise that the French metaphysician had not the slightest interest in the actual achievements of the empiriometric enterprise, and was eager to dismiss the subject as quickly as possible. It suited him to reject modern science outright as a "*savoir ignorant,*" a misguided pursuit which has nothing positive, nothing of any real value to contribute: "The least that can be said," he tells us, "is that the whole business is rather pointless." As Jean Borella notes: "For him it is only one production among others of a world that he condemns *en bloc.*"

On top of it all, Guénon denies even the originality of this disdained science: "The profane sciences," he writes, "of which the modern world is so proud, are really and truly only degenerate 'residues' of the ancient traditional sciences."[3] Yet, far-fetched as this claim appears (and in fact proves to be), Guénon has his eye on a major metaphysical truth; for he goes on to say: "just as quantity itself . . . is no more than the 'residue' of an existence emptied of

3. Ibid., p.5.

everything that constituted its essence." As we shall have occasion to observe, therein indeed—in this very recognition—lies the key to the metaphysical understanding of what contemporary physics is about. However, it is to be noted that the "residual" nature of quantity does not, by any means, entail that sciences concerned with the quantitative aspects of reality are themselves, in a way, "residual." To substantiate that modern physics, in particular, constitutes a "degenerate residue" of some traditional science, one would need to establish that the *modus operandi* of the former derives from an ancient source. To be sure, there exists a certain historical continuity between modern and ancient sciences; Galileo and Newton, for example, were steeped in the Aristotelian tradition, a circumstance which of course played a role in the development of their thought. Yet, even so, the decisive fact is that they *broke* with the physics of Aristotle, and replaced it with something else, something that is new. The *modus operandi* of Newtonian physics, in particular, is most assuredly *not* derived from the Aristotelian tradition, much less does it stem from any authentically traditional science. And what I find most ironic: if it *had* been thus derived, Guénon would, by this very fact, *not* have dismissed that physics as a profane science, bereft of all cognitive value!

It is strange, too, that Guénon should deny the originality of the contemporary scientific enterprise, given his belief that the actualization of generic possibilities is linked to the successive phases of a major cycle or *Manvantara*. In keeping with that doctrine, it would appear that modern science constitutes precisely the cognitive possibility that is "strictly in conformity with the conditions of the cyclic phase at which humanity has now arrived," to put it in Guénon's own words; and this means not merely that it constitutes, basically, the only *viable* kind of science in the present age, but also, and by the same token, that it is a "way of knowing" which could *not* have been effectively pursued in bygone times. This explains, moreover, why modem science and its technology constitute in fact the one domain in which our civilization clearly excels all others, and exhibits a kind of mastery not to be found in the ancient world. The intentional object of contemporary physics may indeed be a "residue," metaphysically speaking, but the science itself is very far from being such.

The decisive event in the evolution of modern thought was no doubt the exclusion of essences effected by Galileo and Descartes, and the concomitant adoption of a bifurcationist epistemology which relegates perceptible qualities to the subjective domain. These metaphysical and epistemological infractions, however, do not in themselves invalidate the *modus operandi* of a science concerned exclusively with the quantitative aspects of reality. From a methodological point of view, the exclusion of essences constitutes simply the delimitation that defines and thus constitutes the domain of physical science; and it is by no means paradoxical that the science in question owes its prowess precisely to that very reduction of its scope; as Goethe has wisely observed: "*In der Beschränkung zeigt sich der Meister.*"[4] Let us note, at the same time, that since the logic of contemporary physics is positivistic or operational, as the prevailing philosophies of science aver, that science has nothing to do—on a technical plane!—with the Cartesian premises; and if it happens that contemporary physicists, in their scientistic beliefs, remain affected by a residual Cartesianism, this does nothing to invalidate the positive findings of physics as such. The knowledge in question may be miniscule by comparison to higher modes, and may indeed conduce to dissolution, as Guénon avers, but constitutes, even so, a *bona fide* though partial mode of knowing.

On the other hand, Guénon's failure to distinguish between science and what he terms "scientific mythology" does not invalidate his perception of the scientific enterprise as the dominant factor driving contemporary humanity "downwards" to the end-point of its cycle. He broaches the question by pointing out that the public at large is prone to accept "these illusory theories" blindly as veritable dogmas "by virtue of the fact that they call themselves 'scientific,'" and goes on to note that the term "dogma" is indeed appropriate, "for it is a question of something which, in accordance with the anti-traditional modern spirit, must oppose and be substituted for religious dogmas."[5] What follows, in *The Reign of Quantity,* is an elaborate analysis of the modern and indeed postmodern

---

4. Literally translated: "In delimitation the master shows himself."
5. Ibid., p.121.

world, which has rarely, if ever, been equaled either in depth or in breadth.[6]

It is of major importance to recall that Guénon distinguishes two principal phases in the ongoing descent, which he designates by the terms "solidification" and "dissolution"; and it is of interest to note that he enunciated this distinction at a time when physics was just entering the second aforesaid phase through the discovery of quantum mechanics. Although Guénon displayed no more interest in the new physics (which came to birth between 1925 and 1927) than in its Newtonian predecessor, and seems hardly to take note of the quantum revolution, it is clear that the advent of quantum theory does indeed mark the de-solidification of the physical universe. Not only, however, does this development—which came as a complete surprise and major shock to the scientific community—accord with the principles of the Guénonian analysis, but as I will show in the sequel, that analysis provides in fact the key to a *metaphysical* understanding of quantum theory, and thus of contemporary physics at large: the very science, that is, the existence of which Guénon never recognized!

What we propose to do is to complement the Guénonian critique by considering quantum theory, in particular, from a traditional metaphysical point of vantage, in accordance with the teachings of Guénon himself.

Ages before the advent of modern science, human knowing began its fated descent. All of recorded history corresponds already to an advanced stage of the decline to which St. Paul refers as a "darkening of the heart": a darkening, that is, of the *intellect*, properly so

6. One can only, in retrospect, lament that the Catholic authorities did not pay heed to that critique when Guénon was writing and lecturing in their midst, and that, instead of taking to heart *The Crisis of the Modern World* (which first appeared in 1927), they became enamored with Jacques Maritain's *Integral Humanism*. How different the subsequent history of the Church might have been if its intellectual leaders had listened to René Guénon! But they did not; and in place of a metaphysically-based critique of scientism have presented us, sad to say, with the likes of *Gaudium et Spes*.

called. It needs however to be understood that in this ongoing descent, the advent of physical science marks a discontinuity, the commencement of a new phase. Prior to this, all human knowing was yet directed towards essence: the essential as distinguished from the material pole of existence. Thus—whether we realize it or not— even the humblest act of cognitive sense perception involves an intellectual apprehension of essence. To be sure, that apprehension has become obscured in varying degrees; yet even so, perception hinges, now as before, upon a discernment of essence: it is this, precisely, that renders the perceptual act *cognitive*. However, with the advent of modern physics the picture has changed: for the first time ever, man's gaze could be directed *downwards*, away from the pole of Essence, towards the *materia secunda* that bounds our world on its nether side. A new methodology, a brand new way of knowing, was needed to accomplish this feat, and came in fact to be inaugurated by the pioneers of the empiriometric enterprise. The first decisive steps were taken in quick succession by Galileo, Descartes, and Newton; then followed two centuries of intense activity—which witnessed, among other things, the discovery of electromagnetic fields and Einsteinian relativity—and then, around 1925, the new physics came finally into its own with the discovery of quantum theory: at long last, essence had now become fully exorcised from the so-called physical universe. To the astonishment and indeed the chagrin of the scientific community, that universe became thus "desolidified"; as Arthur Eddington was quick to observe: "The concept of substance has disappeared from fundamental physics."[7] After more than two centuries of concerted endeavor, empiriometric science had finally attained to "the 'residue' of an existence emptied of everything that constituted its essence," to put it in the words of René Guénon.

What confronts us here is evidently a strange and indeed unprecedented way of knowing. One knows the mass of the electron, its charge and magnetic moment; one knows with consummate precision how it responds to electromagnetic fields, and can utilize electron beams to transmit text or pictures to a fluorescent screen; yet,

---

7. *The Philosophy of Physical Science* (Cambridge University Press, 1939), p. 110.

when asked "What *is* an electron?" one has not the ghost of an idea. It could not be otherwise: for if, indeed, the object in question be "void of essence," then it *has no* quiddity, no "whatness" or *Sosein* at all. Now, as it happens, this curious state of affairs was recognized early on by the founders of quantum theory. Werner Heisenberg, for one, has pointed out that these so-called quantum particles constitute what he termed "a strange new entity midway between possibility and reality," which in a way represent what he termed "a quantitative version of the old concept of '*potentia*' in Aristotelian philosophy"[8]; and Erwin Schrödinger notes that "we have been compelled to dismiss the idea that such a particle is an individual entity which in principle retains its 'sameness' forever. Quite to the contrary, we are now obliged to assert that the ultimate constituents of matter have no 'sameness' at all." And he goes on to press the point:

> And I beg to emphasize this, and I beg you to believe it: It is not a question of our being able to ascertain the identity in some instances and not being able to do so in others. It is really beyond doubt that the question of 'sameness', of identity, really and truly has no meaning.[9]

It *can* have no meaning, let us add, precisely because these putative particles are void of essence: it is essence, after all, that bestows unity, sameness, or identity. In the absence of "unity, sameness, or identity," however, one cannot speak of *being*: nothing that lacks essence, therefore, can exist as an entity, as a being or "thing." The physical universe, conceived as an aggregate of so-called quantum particles, constitutes thus a "sub-existential" domain that needs to be distinguished categorically from the corporeal, as I have pointed out repeatedly.

The objection is sure to be raised that corporeal objects, being composed of atoms, do in fact constitute aggregates of quantum particles; and it is to be noted that even Heisenberg and Schrödinger, notwithstanding their penetrating insight into the nature of these

8. *Physics and Philosophy* (New York: Harper & Row, 1962), p. 41.
9. *Science and Humanism* (Cambridge University Press, 1951), p. 18.

particles, were yet of that belief. Whereas single particles are admittedly "sub-existential," it is claimed that sufficiently large aggregates are not; somehow the sheer number of constituent particles, or size of the aggregate, is supposed to bestow *being*. The ontological distinction between the corporeal and the physical domains is thereby denied, which is to say that the corporeal is reduced to the physical, as almost everyone today believes.

Now, as I have shown elsewhere, the very possibility of a mathematical physics is based upon the fact that every corporeal object X is associated with a corresponding physical object SX, which in the final count reduces to an aggregate of quantum particles.[10] The crucial point, however, is that X and SX are not identical, that in fact they belong to different ontological planes, different intentional domains. What differentiates the two, of course, is the intrusion of essence, or of substantial form, on the corporeal plane: it is this additional component that bestows unity, sameness, or self-identity to corporeal entities, qualities which SX as such does not possess. A distinction needs therefore to be made between atoms in X and atoms in SX, the point being that within a corporeal entity, the very atoms and molecules partake somewhat of essence: of the substantial form, namely, which constitutes the very being of that entity. Thus they become *more* than atoms and molecules as conceived by the physicist: as components of X they constitute genuine *parts* of a whole. Thus conceived, they no longer pertain to the quantitative order: as partakers of essence—even in their capacity as parts—they are no longer mere quantities, no longer physical entities in the strict contemporary sense. Thus, in conceiving the molecular constitution of a corporeal object X as a mere aggregate SX of quantum particles, something essential—quite literally!—has been lost: one is left in truth with a mere residue of an existence "emptied of everything that constituted its essence" precisely as Guénon declares.

As a rule, SX determines the *quantitative* properties of X; and this is the reason, of course, why there can be a mathematical physics.

10. *The Quantum Enigma* (Tacoma, WA: Angelico Press/Sophia Perennis, 2012), chap. 2.

The physical and chemical properties of salt, for example, can be accurately deduced from the physics of NaCl molecules. However, the possibility of deviations exists, and is no doubt realized in varying degrees as one moves upwards along the *scala naturae*; the fact that within a corporeal being the very atoms which constitute its material basis "partake somewhat of essence" is not without consequence even in a measurable or quantitative sense. The reductionist premise may have had its use as a heuristic hypothesis, but sooner or later turns counterproductive; there is reason to believe that in such fields as medicine and pharmacology, for instance, a non-reductionist outlook could open the door to deeper levels of research. What is called for, clearly, is the categorical distinction between X and SX, a recognition that is long overdue.

This brings us to a second major point which the physicists have missed. It is to be noted that the "receptivity" of quantum particles to *essence*—their capacity, namely, to become authentic parts of a whole—is due precisely to the so-called "indeterminacy" characteristic of the quantum domain. If these particles did possess "sameness," did possess self-identity, they could not be thus amalgamated into a corporeal entity. For this to occur, the particles must partake of *potency* in the Aristotelian sense, a qualification which manifests itself to the eye of the physicist precisely as so-called quantum indeterminacy. What the physics community has regarded as an anomaly bordering upon paradox, and what Albert Einstein decried as unthinkable, turns out to be a metaphysical necessity.

No one on either side of the Copenhagen debate seems to have realized that the role of quantum particles is not to bestow, but to *receive* being. To be sure, Heisenberg did refer to these so-called particles as *potentiae*; however, in so doing, he was thinking exclusively of measurement as the process whereby these *potentiae* are actualized. Apparently it did not occur to him that corporeal being as such, so far from reducing to an aggregate of quantum particles, constitutes an actualization—a passage from potency to act—which is to say that SX is actualized in X. It is to be noted, moreover, that even from a scientific point of view this is no mean fact; for indeed, it explains, for example, why billiard balls do not bilocate, and why cats cannot be both dead and alive, as I have pointed out elsewhere.[11] One might add that from a traditional perspective, the passage from SX to X constitutes indeed an act of measurement, though obviously of a kind unknown to the physicist—a point to which I shall return.

These observations may suffice to bring home a major fact: it turns out that the true significance of contemporary physics can only be discerned from an authentically *metaphysical* point of vantage. Only thus does the new physics become philosophically comprehensible: only thus does it cease to be what Whitehead has characterized as "a kind of mystic chant over an unintelligible universe."

Given that contemporary physics deals ultimately with entities pertaining to a sub-existential plane, it behooves us to consider, at least briefly, how knowledge of that kind can be attained. Whereas mankind has always been in possession of means by which *higher* realities could be known, it was apparently left to the twentieth century to discover a way of knowing things that do not truly exist. The question is: how does one accomplish that prodigy?

To understand the logic of contemporary physics, we need first of all to distinguish between its laws and what I will provisionally call physical entities. There is much to be said in support of the view, first enunciated by Eddington, that the fundamental laws of physics (including its universal dimensionless constants) can be deduced from the *modus operandi* of physical science, which is to say that these laws pertain to mathematical structures imposed by the physicist himself through mensuration. I will only add that this claim has been strikingly confirmed in recent times by the American physicist Roy Frieden, who has in fact deduced the laws in question through an information-theoretic analysis of the corresponding instruments of measurement.[12] What I have termed a physical entity, on the

11. "From Schrödinger's Cat to Thomistic Ontology," *The Thomist*, 63 (1999), pp. 49–63.

12. *Physics from Fisher Information* (Cambridge University Press, 1998).

other hand, is what we actually detect and measure through these instruments. To be sure, a physical entity is in a sense comprised of quantum particles; and yet, what our instruments detect is not, strictly speaking, a particle, or a set of particles, but an associated probability distribution. A strange recognition: what the physicist observes and measures is, finally, none other than a *probability*!

On closer examination, however, the concept of probability proves to be singularly appropriate; as Heisenberg points out, a so-called probability does in a way constitute "a quantitative version of the old concept of 'potentia' in Aristotelian philosophy." A probability, after all, is not itself a "thing," but something that points beyond itself to a "thing or event" of which it is the probability. It would be misleading, therefore, to attribute "existence" to a probability; but neither can it be said that a probability is simply nothing at all. Thus it is indeed "just in the middle between possibility and reality," exactly as Heisenberg maintains.

The notion of a probability, of course, did not originate in a quantum-mechanical context. Long before the quantum era, mathematicians were calculating probabilities associated with such mundane phenomena as tossing a coin or dealing a hand from a deck of cards. What is new is the idea of probability as a kind of physical entity, and in fact, fundamentally, as the *only* kind. Yet the underlying mathematical conception remains the same: a probability, thus, is simply a *weighted* possibility. It is here, in this qualification, that quantity enters the picture; and it enters, not (like in classical physics) as an attribute of a corporeal existent (say the mass or diameter of a solid sphere), but as a measure of likelihood. That such "measures of likelihood" should somehow subsist in the physical universe and propagate as waves—that in fact they are precisely what *does* "exist" in that universe!—this, to be sure, comes as a surprise. It is something, moreover, that the physicist as such is categorically unable to interpret, unable to grasp. What is called for, clearly, is a *metaphysics*. It turns out that, quite unexpectedly, the physicist is catching a glimpse of *materia*, of the Aristotelian *hyle*. Not in itself—not as a "pure potency" or a mere possibility—but as a *weighted* possibility: as a *probability*, to be exact. Whether he realizes it or not, the quantum physicist is looking in—through a keyhole, as it were—

at the mystery of cosmogenesis: not in the bogus sense of big bang theory, but *ontologically*, in the here and now. By way of quantum theory he has entered upon an ontological domain "prior" to the union of matter and form: *onto a sub-existential plane which presumably has never before been accessed by man.*

How, then, does the physicist gain knowledge of the probability distributions which "inhabit" and in a sense constitute that uncanny domain? Basically, he knows them the way probabilities have always been known: by calculation, namely, or by sampling. Consider, for instance, the probability of getting "three of a kind" when five cards are chosen at random from a deck: one can calculate that probability, or deal a thousand poker hands, count the number of times the given event occurs, and divide by a thousand. In the case of physics, calculation is of course effected by way of fundamental laws and universal constants, whereas sampling is carried out by measurement. What needs to be emphasized is that a quantum-mechanical probability distribution can be "sampled," can thus be "observed," precisely because the probabilities in question refer to things or events pertaining to the *corporeal* plane. The fact, moreover, that these probabilities can be approached from two directions—by way of theory and by way of measurement—is precisely what opens the door to a positivistic or operational kind of knowing: the Baconian kind, namely, that powers our technology.

Yet man was born to know, not quantities, but essences. He is unable, in fact, to think of quantity by itself, without reference to substance; and if the very concept of substance has indeed been exorcised from contemporary physics, he is bound to reintroduce it in one way or another, and as one might say, by way of the back door. It is humanly impossible to think *in abstracto*, without somehow reifying the "things" we conceive. Now, on a technical plane such reification is permissible as an artifice, a means that permits us to "think the unthinkable" if you will; and this is in fact how a mathematician, for example, does think of such things as n-dimensional spaces and other abstract and unimaginable structures. What saves him from error, from a kind of intellectual idolatry, is the recognition that the images he fashions in his mind—what the Scholastics termed *phantasmata*—are no more than stepping stones, what the Germans call "*eine Eselsbrücke*" (a "donkey's bridge"). When it comes to theoretical physics, on the other hand, such an attitude can hardly be maintained; it belongs to the very definition of physics, after all, to conceive of its intentional objects as real or existent things. To be sure, the physicist can grasp the idea of a probability, for example, without illicit reification—but only because probabilities refer to actual things and events. The moment one denies the substantial reality of these "things and events," on the other hand, one has created an intellectual vacuum that cannot be sustained; one can no more endure such a vacuum than live without breathing. Imperiously the void must be filled by an attribution of reality, some stipulation of being: there is no such thing as a *Weltanschauung* without an ascription of substance. And this leaves the contemporary physicist with two options: he can attribute substance to substantial things, and thus rediscover the corporeal world, or he can posit substance in the physical domain, where the very notion does not apply. These are, basically, his only choices, and no amount of mental acrobatics can alter this fact.

Curiously enough, physicists are invariably loath to acknowledge the reality of the corporeal world. For some reason they cannot bring themselves to acknowledge such a thing as color: the fact, for example, that red apples are red—which in truth is all it would take to affirm the corporeal order. And thus they condemn themselves to a state of chronic schizophrenia, for it goes without saying that everyone, in his daily life, does believe staunchly in such things as red apples. Moreover, by virtue of the aforesaid alternative the physicist is obliged somehow to smuggle the notion of substance into a sub-existential domain, in which *by his own canons* substance has no place. For over seventy years, now, some of the brightest scientific minds have applied their ingenuity in this unpromising pursuit, and in the process have created what may indeed be the most fantastic literature the world has ever seen. One can choose today among different varieties of "many-worlds" theory, or if one prefers, can find comfort in the idea that "Observers are necessary to bring the Universe into existence," as the so-called Participatory Anthropic Principle affirms.[13] The neutral bystander cannot but ask what it might be that drives intelligent men to engage in this curious enterprise, this—dare we say it?—*this madness*. The question has of course no easy answer. What confronts us here is not a fringe phenomenon—not the conduct of amateurs or lunatics—but the unfolding of tendencies and ideas indigenous to the physics community. As I have argued elsewhere, contemporary physics, in its highest theoretical formulations, is presently seeking to transform itself into a hyperphysics: a kind of mathematical metaphysics or theology, one can almost say.[14]

The phenomenon, I believe, can only be understood in basically Guénonian terms. We seem to be witnessing, at least in its initial phase, the self-destruction of mathematical physics, the inevitable *reductio ad absurdum* of a science bent upon "quantity itself." It appears that, in the end, the physicist—I mean the theoretical as opposed to the applied physicist—is driven to engage in the construction of formal worlds in a Promethean endeavor to arrive at a *total* understanding of the universe. The very tendency which has led, at an earlier stage, to the creation of a *bona fide* physics eventually goads him on to overreach, and thereby to dissolve the former in a hyperphysics, a pseudo-science that has lost touch with physical reality. In a sense René Guénon may be right in his pessimistic evaluation of physics; the crucial point, however, is that the building of a hyperphysics constitutes a new phase in the evolution of physical science: the phase of decline, namely, of eventual termination. In the final count, nothing that is not centered upon essence, and thus ultimately upon God, can avoid dissolution, dispersion into nothingness. Here too, it seems, the words of Christ apply: "He that gathereth not with Me scattereth abroad." The flight from Essence cannot but lead finally to "outer darkness."

13. I have discussed these matters at some length in *The Wisdom of Ancient Cosmology*, (Oakton, VA: Foundation for Traditional Studies, 2003), chap. 11.

14. This goal has meanwhile been realized by Stephen Hawking in *The Grand Design*.

The foregoing analysis has brought to light the underlying cause of scientistic illusion. Why does a penumbra of scientistic belief accompany in practice even the highest flights of scientific insight? Or equivalently: why is the physicist inevitably driven to transgress, in one way or another, the canons of physics itself? For exactly the same reason, it turns out, why he cannot bring himself to admit that red apples are red: the cause of this strange phenomenon is the rejection of essence, and thus the de-essentialization of the world. I mean, of course, the *corporeal* world, the one in which we find ourselves: the only world which, in our present state, we can experience, can know "existentially." In the final count, it is this de-essentialization of the corporeal that forces the physicist to stipulate substance where substance has no place, and thereby obliges him to succumb to scientistic illusion. Once the act of de-essentialization has been perpetrated—once "God has been slain," to put it in Nietzschean terms—the next step becomes inevitable: something void of essence must be unwittingly essentialized: a false god, if you will, must be installed in place of the true. *Scientism proves in the end to be the idolatry of a post-Christian civilization.*

Science, on the other hand, is something altogether different: as different as methodology is from metaphysics. Whereas, metaphysically, de-essentialization constitutes the fundamental error that begets scientism, methodologically it constitutes (as we have previously noted) the delimitation that renders possible a new way of knowing: a science in which mathematical symbols replace essences, and our gaze is diverted from the external world to a domain of ciphers, the meaning of which is defined in operational terms. As Francis Bacon had shrewdly foretold, it is indeed a knowing accomplished through a *novum organum*—a kind of machine for the mind—a knowing in which truth and utility become in effect identified.

However, in addition to its function as a purveyor of positivistic truth, modern physics admits of a hermeneutic which escapes the Baconian reach. Truth, however positivistic or operational it may be, is yet something more than utility: even if it can be measured, so to speak, in terms of prediction and control, it cannot be thus defined. If "words derive their meaning from the Word," as Meister Eckhart declares, must not truths likewise derive their veracity from Truth? The Baconian reduction of truth to utility corresponds, epistemologically, to the de-essentialization of the world: here too, in the realm of knowing, the essential has been cast out. And yet the essential *cannot* in fact be cast out: what would remain would be nothing at all.

There must, consequently, be another side even to contemporary physics, a side which only the metaphysician can perceive. If it be his task to inveigh against a spurious mythology promulgated in the name of a positivistic science, it behooves him even more to uncover the metaphysical significance of its actual findings; as Seyyed Hossein Nasr has made amply clear in his Gifford Lectures, the failure to integrate science into higher orders of knowledge is indeed fraught with tragic results for humanity.[15] How then, let us ask, can such an integration be accomplished?

It is needful, clearly, to begin with the foundational science, which can be none other than quantum theory; and here we have already accomplished the essential. The first and crucial step consists perforce in the ontological discrimination between the corporeal and the physical domains. It is this metaphysical discernment that situates the physical universe, properly so called, as a subcorporeal plane, and thereby integrates the intentional object of contemporary physics into the traditional ontological hierarchy. That same recognition, moreover, enables us to understand what quantum theory—and thus contemporary physics at large—is actually about. It permits us, as we have seen, to interpret the phenomenon of quantum indeterminacy ontologically, and thereby to comprehend the nature and function of quantum particles from a metaphysical point of view. Richard Feynman once remarked that "no one understands quantum theory," and this is in a way correct: no one *can* understand quantum theory *philosophically* without distinguishing between the physical and the corporeal planes.

15. *Knowledge and the Sacred* (New York: Crossroad, 1981), p. 207.

Consider the phenomenon of state vector collapse, which has mystified physicists since the Solvay Conference in 1927. So long as the corporeal is reduced to the physical and the two domains are thus confounded, this quantum mechanical phenomenon does remain truly inexplicable. What stands at issue is the fact that the interaction of a physical system with a measuring instrument results in a determination, a reduction of indeterminacy, for which there is no *physical* explanation. The moment it is realized, however, that the measuring instrument is perforce corporeal, it becomes clear that the determination in question constitutes an act of essence, and thus of form. It is in fact the very nature of essence, of form in the Aristotelian sense, to impose bounds: and that is why manifestation, the union of matter and form, has been traditionally conceived as an act of measurement. As Ananda Coomaraswamy explains:

> The Platonic and Neoplatonic concept of measure agrees with the Indian concept: the 'non-measured' is that which has not yet been defined; the 'measured' is the defined or finite content of the universe, that is, of the 'ordered' universe; the 'non-measurable' is the Infinite, which is the source of both the indefinite and of the finite, and remains unaffected by the definition of whatever is definable.[16]

Now, that which measures is not the non-measured, but form, which is precisely an act. The physicist, therefore, has every right to be perplexed: *the act of determination constituting state vector collapse cannot be explained in terms of the physical, which is void of essence, void of substantial form.* That collapse is therefore indicative of a nonphysical cause, a principle that comes into play on the corporeal plane.[17] Whether he knows it or not, by way of state vector collapse the physicist has in truth detected the cosmogenetic act. Having penetrated beneath the *terra firma* of our world to the level of the "waters" below, which remain even after the Spirit of God has

---

16. Quoted by René Guénon, op. cit., p. 37.

17. On the subject of state vector collapse, I refer to *The Quantum Enigma*, op. cit., chap. 3.

"breathed upon their surface," and having captured in probabilistic terms something of the primordial chaos, the *tohu-va-bohu* of that sub-existential realm, the physicist re-enters the corporeal plane through the act of measurement, and in so doing witnesses the alchemical marriage of matter and form. Now, this recognition of what quantum physics is actually about constitutes a step in that integration of modem science into higher orders of knowledge to which Professor Nasr alludes as a major desideratum.

I will mention that in addition to state vector collapse, quantum physics has presented us with a second seeming absurdity: the phenomenon, namely, of *non-locality*. It appears that the quantum world is knit together more closely than the canons of the Einsteinian space-time continuum allow, implying that the cosmos in its integrality does not actually fit the confines of that continuum. As I have pointed out in an article on Bell's theorem,[18] this discovery is tantamount to a recognition of the intermediary, or what occultists term the astral, domain—the *bhuvar* of the Vedic *tribhuvana*—which has not only been excluded from the purview of modem science, but has long ago faded from the horizon of Western cosmology. Curiously enough, quantum mechanics takes us not only downwards, beneath the *terra firma* of the corporeal world, but also, it appears, in the opposite direction: "upwards," beyond the spatiotemporal order, into the astral plane. Quantum physicist Henry Stapp may be right in referring to non-locality as "the most profound discovery of science": from an ontological point of view it is indeed the most profound in its implications. One may surmise that St. Thomas Aquinas, for one, would have been fascinated by this discovery, and might have composed, at the very least, an *opusculum* to explain its ontological bearing.

It is of interest to recall that despite his radically negative appraisal of modern science, René Guénon himself was not averse to the idea

18. Reprinted, with some improvements, as Chapter 4 in *The Wisdom of Ancient Cosmology*, op. cit.

of a mathematical exemplarism: the notion that a mathematical structure can point beyond itself to a metaphysical reality. Thus, in his monograph on the infinitesimal calculus, after reflecting at length on the formation of a mathematical integral, he concludes:

> These indications show that the things in question are capable of receiving, through an appropriate analogical transposition, an incomparably larger sense than they seem to possess in themselves, since, by virtue of such a transposition, integration and the other operations of the same kind appear truly as a symbol of metaphysical "realization" itself.[19]

Surely, a more splendid example of mathematical exemplarism can hardly be conceived! It is consequently all the more surprising that Guénon showed so very little interest in mathematical physics, and contented himself in that domain with a patently superficial analysis: an account that fails to distinguish between physics, properly so called, and that "scientific mythology" with which it is confused in the popular imagination. On the other hand, it is only by way of quantum theory that the foundational logic of physics has at last come to light, a development regarding which Guénon appears to have been insufficiently informed. The probabilistic mode of knowing, in any case, was evidently foreign to the great metaphysician, whose conception of modern physics seems to have remained "classical" to the end. In a word, Guénon lacked the means to comprehend the *modus operandi* of physical knowing—how the physicist can know things that do not actually exist—and was consequently predisposed to conclude that in fact the latter does not know at all.

---

19. *The Metaphysical Principles of the Infinitesimal Calculus* (San Rafael, CA: Sophia Perennis, 2004), p. 118.

3

SCIENCE AND EPISTEMIC CLOSURE

IN THE PRECEDING CHAPTER we have been concerned with the so-called "de-essentialization" of the world wrought by the empirometric enterprise, and have uncovered a hidden nexus between *bona fide* science and what may be termed scientistic belief. We propose now to consider that "de-essentialization" and the concomitant emergence of scientistic belief in light of the French philosopher Jean Borella's magisterial study of symbolism.[1] Centered specifically upon language and thought as such, Borella's opus embodies a philosophical point of view that enables one to survey the aforesaid phenomena in terms of a single conception—what he terms *"epistemic closure"*—which at one stroke lays bare the imperative of "de-essentialization" and of that "nexus between science and myth" which stands at the center of our quest. In the first part of this chapter I will place before the reader the aforesaid conception, following which I propose to consider some of its major implications for the philosophy of science, with special reference to modern physics.

It may be well to begin with a few words concerning the distinguished French scholar whose doctrine has inspired the present chapter. A born philosopher, he has himself characterized his bent as "instinctively Platonist." By the time he became acquainted with the writings of René Guénon during his college years, he perceived the Guénonian doctrine as an exposition of the Platonist metaphysics

---

1. Major portions of that study have been published in the following books: *Histoire et théorie du symbole* (L'Age d'Homme, 2004), *La crise du symbolisme religieux* (L'Age d'Homme, 1990), and *Penser l'analogie* (Ad Solem, 2000).

"such as I discovered in myself." Due perhaps to the influence of Guénon, the young philosopher acquired an intimate knowledge of Eastern metaphysical doctrines, without however becoming alienated from his Western roots: these too he discovered and embraced, as he himself relates:

> I went back to ancient doctrines like a delighted child going from discovery to discovery, from treasure to treasure, from marvel to marvel. I recognized and loved this Christian past, its beauty not unworthy of the God whom it had honored with its liturgy, cathedrals and theologies. It was in me as flesh of my flesh, soul of my soul, heart of my heart, and I did not know it. Once discovered, fixing the gaze of my spirit upon the holy Fathers and Doctors, upon the Clements, the Dionysii, the Gregories, the Augustines and the Thomases, I said: I too am of their race. Surely not by sanctity or genius, but by blood. Drinking in the freshness of the ages, I felt my Christian soul revive. . . .[2]

It is evident from this remarkable profession that philosophy and theology could not but be inseparably linked in the thought and writings of this Christian Platonist. Thus it was in part the controversy resulting from the proclamation of the Bodily Assumption of Mary, the Mother of God, promulgated in 1950 by Pope Pius XII, which motivated Borella's doctoral dissertation in philosophy. As Borella explains, it was the almost universal disbelief and incomprehension with which this dogma was met even in Catholic intellectual circles that

> elicited from me what seemed to be a self-evident response: beyond the divisions and oppositions of analytic reason stands the truth of the real, one with itself, inseparably both historical and symbolic, visible and invisible, physical and semantic. This self-evident response rested upon a kind of direct and sudden intuition in which was revealed, obscurely but without any possible doubt, the ontologically spiritual nature of the

2. *La charité profanée* (Editions du Cèdre, 1979), p. 32.

matter of bodies, without for all that casting any doubt on the reality of their corporeity.[3]

I find these words to be singularly profound, and suggestive of a philosophy which is at once rational, in the highest sense, and nonetheless authentically "mystical" as well. The more deeply one probes Borella's philosophical doctrine, the more one senses that its essential content derives from a single presiding insight which could only have been given in a "direct and sudden intuition," a kind of gnosis, one can say. From the start Borella has situated himself "beyond the divisions and oppositions of analytic reason," on the unshakable ground of intellective vision. He is acutely cognizant of the fact that, *on a mental plane*, vision is perforce mediated by concepts, even as, on a sensory level, it is mediated by signs and symbols, be they natural or culturally instituted. With consummate clarity he recognizes that we see *through* the sign, and therefore behold, not the sign, but indeed its referent. And from the very outset of his philosophical inquires he appears to understand that the enigma of "semanticity" derives in fact from the central mystery of Christianity: the mystery of the *Logos*, the Word that *is* God.

In a remarkable discourse[4] regarding "language and thought," Borella takes as his starting point the provocative dictum of Condillac: "Science is only language well posed." Certainly Borella does not accept the thesis that science is no more than language "*bien faite*"; he does, however, maintain that "one may consider that property as the criterion of scienticity (*scientificité*)." This means that science—at least in the contemporary sense—is characterized by its use of language: by the logic, one can say, of its formal expression. What is it, then, that differentiates the scientific from the pre-scientific use of language?

---

3. *Symbolisme et réalité* (Ad Solem, 1997), quoted from an unpublished translation by G. John Champoux.

4. *Histoire et théorie du symbole*, op. cit., chap. IV, art. I.

Borella begins his analysis by distinguishing categorically between "thought" and language. What is "thought"? It is a mental movement in quest of an object, he replies. Thought, therefore, is inherently oriented towards an objective referent by way of a concept, which Borella defines in Scholastic terms as "the form of an act by which the understanding intends an object." What we think, therefore, is always the object, even though we think the object by means of a concept. What, then, is language? One may characterize language by its function, which is to support, express, and communicate thought. Thought, therefore, has primacy. Having thus distinguished between thought and language, Borella goes on to observe that the question of truth—of coherence and non-contradiction—arises for both, but that the *criteria* of truth appropriate to these respective levels are vastly different. What matters on the plane of thought is what Borella terms "*l'ouverture à l'être,*" that is, "openness to being." The term of thought, the fulfillment of its quest, resides in the being of its objective referent; for thought it is the transcendent object itself that counts *de jure*. And this fundamental fact entails that the corresponding criteria of truth or coherence are *ontological* and *interior*: "*Verum index sui*" says Spinoza. The case of language, on the other hand, is very much the opposite: here the criteria are perforce *formal* and *exterior*. A key recognition! For as Borella goes on to note: "There results a kind of inverse relation between the coherence of language and of thought. In effect, the more open the thought is to being, the less assured it is of the pertinence of its discourse and the more the latter appears to it as inadequate." This decisive insight calls to mind the last didactic utterance of Aquinas: the "*mihi ut palea videtur*" ("to me appears like straw"), which points beyond the bounds of his "official" doctrine.[5]

What presently concerns us, however, is something more specific: it follows, namely, from the aforesaid principle, that what we are wont to term "scientific exactitude" is to be purchased at a price. What is that price? It is none other than what Borella terms "*the*

---

5. One might add that unfortunately this "didactic utterance" appears for the most part to have been roundly ignored by the latter-day disciples of the Saint.

epistemic closure of the concept," which consists in the elimination from the concept of everything that proves recalcitrant to linguistic or formal expression, and is "epistemic" in the sense that it characterizes the nature of scientific thought. What stands at issue is not in fact a reduction of the concept to language (which is impossible), but a renunciation, on the part of the scientist, of any knowledge concerning the essence of things. What the scientist relinquishes by virtue of epistemic closure is thus precisely the kind of knowing proper to philosophy as such; for indeed, what the philosopher seeks is a revelation of essence in "an illuminative encounter with the being itself of the object," to put it in Borella's words. On the other hand, what the philosopher, for his part, renounces—in a kind of "speculative humility"—is in fact every conceivable closure of the concept in the face of its object; it appears that Whitehead was speaking for philosophy as such when he declared "exactness" to be "a fake." Guided from the start by a supra-rational intuition, which could well be termed a sense of "wonder," the philosopher uses concepts as a means to the attainment of a supra-conceptual truth in a non-discursive act of contemplative vision. As Borella has beautifully expressed it: "Philosophy is love of the divine *Sophia*, that is to say, the self-revelation of the Principle itself; it is the desire for the knowledge by which the Absolute knows itself." Such is the traditional, the *authentic* conception of philosophy: a far cry, obviously, from what academic philosophy has nowadays come to be![6]

Getting back to science, one sees in light of Borella's analysis that there is a principial opposition between science in the contemporary sense and philosophy, properly so called. Not only do the two disciplines tend to different ends, but it happens that the constitutive act of science—the epistemic closure of the concept, namely—is inimical to the philosophic quest.[7] We need now to ask ourselves: what exactly is the end of science, the goal which *de jure* terminates

---

6. We shall return to this question in Chapter 8.

7. One should add that the philosopher is able in his own way to consummate the act of epistemic closure without ceasing to be a philosopher: "The more is capable of the less," as Borella loves to say. If that were not so, there could be no genuine philosophical understanding of science as such.

its search? In answer to this question Borella maintains that science achieves its term precisely in the pragmatic domain, that is to say, in the form of a technology: "There are only, for a living being, two means of ceasing to think: to contemplate, or to act."

Now, these incisive and principial recognitions—terse though they be—suffice to characterize the scientific enterprise in its broadest outlines. The generic effect of epistemic closure, one sees, is to filter out essence, and therefore being. And this means that science is constrained to reduce phenomena to "pure relations," that is to say, relations which are independent of the beings which enter into them. Borella's prime example of such a reduction derives from the physics of Galileo, in which actual bodies are replaced by the fiction of "mass points," between which the relations contemplated by the physicist are deployed. As Borella explains:

> There is thus an identity of nature between the concept and its object since the latter is likewise a concept, whereas in philosophical knowing the concept is only a means by which the object is known: essentially transitive, it remains thus ontologically open. The Galilean universe is therefore a universe of object-concepts which move in a conceived space-time. The geometrization of space entails the disappearance of every qualitative distinction.

What purpose, then, does this Galilean conception—this putative universe—serve? Its epistemic closure renders that notion useless philosophically: the Galilean concept does not lend itself to a knowledge of essences, a knowledge of being. The only possible use, its only feasible and legitimate function, pertains therefore to the sphere of action, that is to say, to what in scientific jargon is termed "prediction and control." Galilean physics conforms thus to the Baconian conception of a science, a way of knowing, if one may call it that, in which truth and utility "are here one and the same," as Bacon himself has put it.

It is to be noted that Borella does not claim to propound a philosophy of science. He makes it clear that more is needed to arrive at such a philosophical doctrine than simply the notion of epistemic closure, which as he points out, is descriptive but not explicative: "It does not suffice to close a concept to produce science." It is moreover to be noted that Borella would be the last to deny the genius of the great founders—from Galileo to Einstein—who have, each in his own way, through a creative stroke constructed an object-concept universe of immense scientific interest. I might mention that Albert Einstein, for one, was cognizant of the fact that these seminal conceptions constitute what he termed "a free creation of the human spirit,"[8] even though he may not have fully recognized the philosophical implications of these foundational "incursions" into the scientific process. Borella speaks in this context of a "legitimate bias," and of a "point of view" which determines the given object-concept universe; but he does not enter into a detailed discussion of these matters. He need not do so: from a strictly philosophical point of view he has stated what in fact is the essential point.

It appears that Borella is primarily interested, not in science *per se*, but in its relation to philosophy. He is concerned, above all, to refute a fatal error: "One supposes today that science is the only form of authentic knowledge, and that the role of philosophy should be limited to the determination, and the description, as accurately as possible, of the different procedures which science puts into effect." His primary task is thus to recover the very idea of philosophy, and demonstrate that there is indeed a *"connaissance philosophique."* That done, the next step is to point out—on authentically philosophic ground!—that science is in principle incapable of understanding the nature of its own findings, for the simple reason that, from its point of view, the epistemic closure upon which it is based remains invisible: "It is only from a philosophic point of view that this circle appears as a circle, that the epistemic closure appears as a closure." It is true, certainly, that all conceptual knowing entails a certain speculative closure; the point, however, is that the philosopher is well aware of this fact: "The philosopher knows that one can only trace an epistemic circle within a wider speculative field: one can limit only with reference to something that is unlimited." It follows that the highest rank in the hierarchy of knowledge belongs

8. *The Evolution of Physics* (NY: Simon and Schuster, 1954), p. 33.

perforce to metaphysics, "since it defines the most general possible speculative field." What primarily concerns Borella are the implications of this decisive principial truth. It follows, first of all, that the vaunted autonomy of the contemporary sciences cannot but be spurious. One has been taught to believe that the individual sciences, in the course of their evolution, have progressively detached themselves from philosophy, and have attained autonomy; and this is true in part: an emancipation from philosophy—a severance of ancient bonds—has indeed taken place. The problem, however, is that there has been a concomitant loss of cognitive content, and an ensuing confusion. As a way of knowing, properly so called, science *cannot* be autonomous; as Borella points out, the only autonomy to which it may attain pertains to the pragmatic realm. Strange as it may seem, it is the *traditional* sciences—the ones we have been taught to regard as "primitive superstitions"—that do have access to authentic knowing by virtue of their connection with philosophy. "The difference between pre-Galilean and post-Galilean science," Borella explains, "is that, under the ancient regime, the boundaries of the different scientific domains within the general speculative field are not entirely closed: the particular sciences remain open to the general science which is philosophy, and which for them is normative."

Though Borella himself does not formulate *"une theorie de la science,"* it happens that his doctrine of epistemic closure furnishes the ideal basis for such a theory. I propose now to pursue that course, far enough to connect with questions relating to the foundations of quantum theory. I will begin with the following observation: *when it comes to the so-called natural sciences, epistemic closure remains perforce incomplete,* which is to say that a discrepancy between the concept and its technical expression is bound to persist. It is only in the case of pure mathematics[9] that the formalization of the

---

9. Inclusive of formal logic, beginning with the meta-mathematical theory of Russell and Whitehead.

concept—namely, its epistemic closure—can actually be effected, which is the reason why, in the case of this science, "we never know what we are talking about, nor whether what we are saying is true," to put it in Bertrand Russell's famous words. When it comes to a science such as physics, on the other hand, we do evidently need to "know what we are talking about," at least to some degree, which entails that epistemic closure cannot be complete. It may indeed be so in regard to the object-concept universe itself; but such a model, all by itself, does not yet constitute a physical science. An auxiliary body of theory, obviously, is required to connect the mathematical model to the empirical realm in which the scientific enterprise receives its validation, and towards which it is oriented; and in this auxiliary technical domain there can evidently be no question of *complete* epistemic closure. Galilean physics, for example, *taken in its entirety*, was far from being epistemically closed; and as a matter of fact, the connection between the Galilean object-concept universe and the corresponding empirical *modus operandi* was, for a very long time, poorly understood. Today it is clear, in light of Einsteinian relativity, that the celebrated "*Eppur si muove*"[10] cannot in truth be validated on rigorous scientific grounds, as Galileo had mistakenly surmised.[11] As Eddington points out: "Relativity theory made the first serious attempt to insist on dealing with the facts themselves. Previously scientists professed profound respect for the 'hard facts of observation'; but it had not occurred to them to ascertain what they are."[12] It is of course to be understood that these "hard facts of observation" do not stand by themselves, but are conceived in relation to the physical theory, and that what is actually "hard" or "rigorous," scientifically speaking, are not indeed "the facts of observation" as such, but the *modus operandi* by which they are connected to the object-concept universe. My point, however, is that this "hardness" or rigor can never be absolute, which is to say (again) that in this auxiliary technical domain epistemic closure

10. Galileo's famous retort affirming the motion of the Earth around the Sun.

11. I have dealt with this question at length in *The Wisdom of Ancient Cosmology* (Oakton, VA: Foundation for Traditional Studies, 2003), chap. 8.

12. *The Philosophy of Physical Science* (Cambridge University Press, 1949), p. 32.

cannot be complete. What confronts us, when it comes to physics as a *total* theory, are *degrees* of epistemic closure; and it appears that the history of science, from Galileo to Einstein and beyond, is marked by successive stages corresponding to progressively higher levels of closure. Strictly speaking, there is no such thing as a "mathematical physics"; what exists, rather, is a physics in process of becoming ever more fully mathematicized.

Whither does this lead? As I have suggested elsewhere, it appears that this evolution is presently entering a new phase characterized by an excessive degree of formalization and a correlative loss of empirical content.[13] A sampling of the contemporary literature in the journals of theoretical physics reveals an abundance of "universe-building" on a scale never heretofore realized. I have argued that eventually physics may cease to be a natural science and turn into what I term a "hyperphysics," a science (or pseudo-science, as one may say) which has lost contact with empirical reality. I am thinking especially of the various "many-worlds" theories which seem to be cropping up these days like mushrooms, or of such a thing as superstring theory, with its object-concept universe of ten or more dimensions (said to collapse, somehow, into the four-dimensional space-time of empirical science). Is this still science, or has it unwittingly turned into science fiction? An unbiased observer can hardly escape the impression that, somewhere along the line, the boundary has actually been crossed, as Richard Feynman, for one, has in fact surmised. It appears that as one approaches the limit of complete epistemic closure, physics becomes—not a "theory of everything" as physicists like to think—but indeed a "theory of nothing at all."

Epistemic closure, as Borella makes clear, entails the elimination of essence, and thus of substance, from the resultant universe. Only at a comparatively late stage in the evolution of modern science, however, did physicists begin to recognize the fact that substance had mysteriously vanished from their world. Eddington was perhaps the first to take note of this "de-essentialization" when he

---

13. *The Wisdom of Ancient Cosmology*, op. cit., pp. 211–215.

declared (in his Tarner Lectures of 1938) that "the concept of substance has disappeared from fundamental physics," a claim which neither Galileo, nor Newton, nor indeed any physicist prior to 1925 has made, or *could* have made. For Eddington, the demise of substance is implied by a remarkable notion, which he apparently was the first to conceive: he maintains that the physical universe is not in truth *discovered*, but rather *constructed* through the *modus operandi* of physics: "The mathematics," he tells us, "is not there until we put it there." What thus distinguishes Eddington's object-concept universe from the Galilean and Newtonian is that the categorical separation between the mathematical model and its operational interpretation has been in principle abolished: in thinking the mathematics, Eddington also thinks, in a formal way, the procedures which have "put it there." The original object-concept universe comes thus to be viewed, not as a model or description of the actual universe, but simply as a mathematical structure defined in operational terms. And let us note that in a physical universe thus conceived, the idea of "substance" has indeed disappeared: such a physics terminates, not in a putative knowledge of objects—of things or substances—but in controlled acts of measurement, and thus, by way of application, in a technology. Eddington claims thus to have carried the formalization of physics to its limit; he claims, in other words, to have enclosed the full body of theory within the epistemic circle by which physics as such is defined.

However, it appears that Eddington may have over-reached: all is not well. According to his "epistemological" analysis, the construction itself—the very procedures by which the mathematics is "put there"—determines not only the fundamental laws of physics, but also its dimensionless constants. For example, Eddington claims to prove—without reference to empirical data—that the so-called fine structure constant is precisely $1/137$; according to the latest measurements, however, this constant turns out to be approximately 0.0072973531, which differs from Eddington's predicted value by about 3 hundredth of a percent. Thought small, this discrepancy—if unresolved—is nonetheless fatal to Eddington's theory: it appears that in his formalization of physics, something must have been left out of account. One is forced to conclude that, after more than four centuries of scientific endeavor, *full* epistemic closure of physics has not yet been achieved.[14]

This brings us to a major recognition: science, in its concrete reality, is not—and cannot be—strictly "scientific." If epistemic closure is indeed the criterion of "*scientificité*," and if in fact this closure cannot be consummated without emasculating the scientific enterprise, then it follows that there can be in practice no such thing as total or absolute scienticity. And therefore, if the outer face of science does conform—by definition, if you will—to the criteria of scientific rigor, there must also be a hidden face which does not. Science too has its "subconscious," which is to say that, in its actuality, it is by no means confined to the "epistemic circle" within which its theory is framed. And this is as it should be; the scientific enterprise, too, must comprise a "dark" component, if one may call it such, which—like the black spot in the white field of the *yin-yang*— plays a legitimate and indeed necessary role in the economy of scientific thought: there could be no creativity, no "flashes of insight" without access to a wider speculative field, which remains unrecognized from a scientific point of view precisely because it is situated outside its epistemic circle. Yet, unacknowledged though it be, that "dark" domain constitutes actually the fertile ground—replete with its imaginative forms, its *phantasmata*—from which those "free creations of the human spirit" are drawn forth by scientists of first rank.

It is to be noted that, in a way, this holds true even in the case of pure mathematics: here too "the dark half of the *yin-yang*" has its role to play. To be more precise: whereas, in the case of mathematics, "perfect rigor" can indeed be attained "at the end" (i.e., in the completed proof of a given theorem), it cannot be consistently

---

14. This does not imply, certainly, that Eddington's theory must be abandoned *in toto*; it means, rather, that the theory needs to be somehow qualified or amended. I should point out, in this connection, that an American physicist by the name of Roy Frieden has apparently succeeded in deducing the fundamental laws of physics from an information-theoretic analysis of the measuring process (*Physics from Fisher Information*, Cambridge University Press, 1995). He does so, however, with the aid of a variational principle which is not itself founded, *à la* Eddington, on epistemological grounds.

maintained in the process of mathematical discovery, be it of theorems or of proof. I am of course referring to the epochal work of Kurt Gödel, known especially for his celebrated Incompleteness Theorem, published in 1931. What Gödel's "meta-mathematical" theorems have brought to light is the fact that it is generally impossible to prove the implications of a given formal system without going *out* of that system. To lay hold of the content of system A, let us say, one requires an enlarged formal system B, and so forth. In a word: *mathematical science, taken as a whole, cannot be fully formalized.* I will mention, in passing, that this absolutely fundamental recognition has decisive implications regarding the nature of "mind" and its relation to neural function, a matter to which we shall return in Chapter 5. What presently concerns us, however, is the fact that Gödel's theorem confirms what we have previously said regarding the limits of scienticity: if not even pure mathematics can be "formalized without residue," what to speak of physics!

There are, in principle, two ways of conceiving the object-concept universe of physics: one may regard it, of course, as an object-concept universe, or one may *reify* that so-called universe, conceive of it, in other words, as "real." To be sure, what differentiates the second "universe" from the first is precisely the attribution of *substance*, a stipulation which, as we have seen, is illegitimate: the idea of substance—a concept that cannot be defined in scientific terms and has no place in scientific discourse—has been introduced spuriously, "smuggled in" so to speak. Let us be clear about it: the Weltanschauung which ensues from that attribution—what Whitehead terms "the fallacy of misplaced concreteness"[15]—is not in truth scientific, but in fact contradicts the very principle of scienticity. As Borella explains, the idea of essence—of being, or of substance—has no place within the epistemic circle to which post-Galilean science, by its very logic, is confined: there can no more be substance within the object-concept universe of modern physics than there can be, let

15. *Science and the Modern World* (NY: Macmillan, 1967), pp. 51–55.

us say, in the Euclidean plane. The reified physical universe—a notion which almost everyone these days seems to accept as an established scientific truth—proves ultimately to be a self-contradiction, on a par with the notion of a square circle.

Now, it has long been my contention that the effects of this fundamental misconception are manifested not only in the individual psyche of the scientist, but likewise in what may be termed the collective psyche of contemporary Western society. As members of that society we find ourselves in a strange predicament: on the one hand we have been conditioned to reify the physical universe, and on the other we continue to believe, now as before, in the "ordinary" world, the familiar universe accessed through sense perception. And although these two universes or worlds are evidently as different as night and day, we are constrained to oscillate between the two, and generally do so, oddly enough, without the slightest scruple or sense of contradiction. As I have argued more than once, the hegemony of science has plunged us into a collective state of schizophrenia from which hardly anyone is able to extricate himself: one moment the grass is green, and the next it is not; one instant bodies are solid, and the next—when we switch our brain to "scientific mode"—they are "atomic aggregates." We seem to be committed to two contradictory world-views: to one on account of our cultural adhesion to the contemporary West, and to the other by virtue of the fact that we are human. It is safe to surmise that just about everyone has been thus afflicted to some degree, in direct proportion, generally, to the amount of education he has received.

What, then, is "scientism": does it reduce simply to Whitehead's "fallacy of misplaced concreteness"? One may of course define "scientism" in terms of that criterion; it should however be noted that the term has other legitimate connotations as well. For example, it can reasonably designate a world-view based upon Cartesian bifurcation, a position which does not necessarily entail the reification of the physical universe: Eddington himself, as a matter of fact, was bifurcationist in his Weltanschauung.[16] I would contend, moreover, that a Darwinist or "evolutionist" world-view is *per se* scientistic,

---

16. *The Wisdom of Ancient Cosmology*, op. cit., chap. 3.

regardless of whether one reifies the physical universe or adopts a bifurcationist epistemology. I am thinking especially of Whitehead, the philosopher who inveighed against "the fallacy of misplaced concreteness" and pioneered the critique of bifurcation, but whose teaching was nonetheless evolutionist to the core, to the point that it provided the inspiration behind so-called "process theology," a doctrine that extends the concept of "evolution" to God himself! There is also "naturalism," an etiological form of scientism, and there is an epistemological version, epitomized in Bertrand Russell's boast: "What science cannot tell us, mankind cannot know." Admittedly, all these scientistic tenets are intimately related, and constitute part and parcel of the contemporary Weltanschauung; yet even so, they are logically distinct and need to be distinguished: that is my point.

Getting back to "scientism" in the first sense—the reification, namely, of the physical universe—let us now ask ourselves how this self-contradictory world-view could have imposed itself upon a major portion of mankind. One might think that the operational validity of physics—the fact that "it works" and gives rise to a miraculous technology—leaves us no choice; but whereas this may be partly true in the case of the uninformed, it can hardly be so when it comes to scientists of first rank. To recognize what ultimately stands at issue, we need to remind ourselves that man was made, not to play positivistic games, but to know truth, to know *being*. It is no more possible for him to renounce the being of things than it is to stop breathing; his hunger for *being*—and indeed, for Being itself, which is God!—is relentless, and cannot finally be appeased by anything less. And so it comes about that when being has been excluded from his purview by an act of epistemic closure, the scientist himself feels compelled to bring it back, to reinstate it somehow in his universe. Admittedly it is possible to obviate the reification of the physical, as we have noted: but only at the cost of locating being in some other domain. It is safe to say that for all but the most wise or the ultra-sophisticated, it will be the physical object-concept universe that is reified, and that the few who manage to avoid this pitfall will likely succumb to some alternative mode of scientism. There is in fact only one way to obviate scientistic illusion, and that is the way of authentic philosophy: one needs to see the whole picture—the epistemic circle *plus* the unlimited speculative field within which it is drawn—in order *not* to be deceived. The moment, therefore, that a science loses contact with what Borella terms "the general science which is philosophy"—in that very instant the birth of illusion is bound to take place. Something alien and indeed contradictory to science is unwittingly smuggled in, and thenceforth masquerades in scientific garb: thus does science of the post-Galilean kind beget scientism. The die has been cast in a radical act of epistemic closure, which cuts the human individual off from true being, or subjectively speaking, from his own true ground and "higher" subconscious.[17]

Now, it is this profound and unobserved schism, I say, that underlies the collective schizophrenia to which we have previously referred, and which in a way "manifests" the aforesaid schism. Severed from his authentic ground, contemporary man has become profoundly disoriented, estranged from the perennial norms. He has thus become vulnerable to the lure of pseudo-norms and specious values which—as if in compensation—contemporary society supplies in abundance. It would be a fatal mistake to suppose that science is neutral in regard to "values," or bereft of ideology as the textbook wisdom declares: nothing could be further from the truth. The fact is that *scientism itself constitutes the ideology of science*, its cultural side, which is in a way a religion, or more accurately: a counter-religion. But these are questions beyond the scope of our immediate concern, which moreover I have dealt with elsewhere.[18]

Following upon these exceedingly general reflections, it behooves us to take a closer look at the object-concept universe of contemporary physics. We know that this universe—the *physical* universe, properly

17. This does not mean that the scientist makes no use of this "subconscious" in the exercise of his scientific functions; as we have noted before, he does, most assuredly, make use of it. My point, rather, is that in the name of epistemic closure the existence and rightful function of that faculty is implicitly denied.

18. See especially *Cosmos and Transcendence* (Tacoma, WA: Angelico Press/ Sophia Perennis, 2012), pp. 141–166.

so called—is supposedly made up of quantum particles; what, then, can one say concerning the nature of these particles? Are they indeed no more than "object-concepts"? Or is it perhaps possible to conceive of them as real entities of some kind?

It is to be noted, in the first place, that these quantum particles and their aggregates are represented in terms of a mathematical formalism; for example, by a so-called state vector in a Hilbert space. Now, the customary or official interpretation of these formal representations is operational, which is to say that the mathematics is interpreted ultimately in terms of an empirical procedure. The meaning of a mathematical formula is thus reduced in the end to an operational statement, a statement of the form: "If you do A, B will ensue," where B is basically the result of a measurement. This is what the experimental physicist is charged to accomplish: his function is to translate the mathematical "statements" of the theoretician into operational terms and puts them to the test.

But the question remains whether operational definition covers the entire ground. One senses that a quantum particle must in truth be *more* than a mere concept, a mere *ens rationis* or "thing of the mind," that it must, in other words, possess a certain objective reality: if it did not, how then could it affect our instruments of detection and measurement? Now, it is true that this question is not in fact scientifically meaningful: it is not what a scientist, *qua* scientist, can ask, let alone answer. To be precise, the condition of epistemic closure, which is the very principle of scienticity, prohibits the scientist from posing that question. As Eddington points out:

> It has come to be the accepted practice in introducing new physical quantities that they shall be regarded as defined by the series of measuring operations and calculations of which they are the result. Those who associate with the result a mental picture of some entity disporting itself in a metaphysical realm of existence do so at their own risk; physics can accept no responsibility for this embellishment.[19]

Like it or not, the idea of substance, of substantial being, has indeed

19. Op. cit., p. 71.

been ruled out by the criteria of scienticity. But whereas scientists accept the notion of scienticity in theory, few if any are able to abide by this condition in practice. It appears that even amongst the most devoted Copenhagenists there is perhaps no one who can accept, fully and consistently, the dictum of Niels Bohr declaring that "There is no quantum world; there is only a quantum description." And rightly so. The intuition that a mere "quantum description" cannot account for a track in a bubble chamber, or for the position of a pointer on a scale, is unquestionably sound. What is not so clear, on the other hand, is whether it is possible to do better: to know *more* than what a merely operational understanding of quantum theory reveals. By what means, in particular, can one gain knowledge of a quantum particle as "an entity disporting itself in a metaphysical realm of existence"? And how, having done so, can one validate such an interpretation, seeing that physics itself "can accept no responsibility for this embellishment"? One can do both, I say, at a single stroke, by availing oneself of what Borella terms "the general science which is philosophy." There can of course be no question of "rigor" with regard to such an interpretation, which is to say that "the general science which is philosophy" is not subject to the condition of scienticity. And that is just the point: the problem at hand cannot be solved "within the epistemic circle" to which science as such is confined. What can and must replace "rigor," in the scientific sense, is a contemplative act of vision, that is to say, an authentically *intellective* as opposed to a merely discursive or mental act.[20]

We need to ask ourselves what kind of an "entity" a so-called quantum particle could be. The question may be put as follows: what is it that we actually measure or register with our instruments? Now, quantum theory itself affirms that what we observe are in fact *probabilities*[21]: not *things*, therefore—not waves, for example, nor indeed particles—but something that is represented mathematically

---

20. It is to be noted that if man were indeed the kind of creature Darwinists take us to be, there could be no such act: under such auspices there could be no intellect, properly so called. As a matter of fact, there could be no mind either, and thus, incidentally, no Darwinists as well.

21. See *The Wisdom of Ancient Cosmology*, op. cit., pp. 63–67.

by a so-called probability distribution. To be sure, probabilities are defined in statistical terms. The idea is simple. The probability of getting "heads" when a coin is tossed turns out to be ½, meaning that, if we toss the coin n times for sufficiently large n, the coin will come up "heads" close to 50% of the time; or to put it more precisely: the deviation from that value will tend to zero as n tends to infinity. The question before us, then, is how a probability, thus defined, may be conceived as "an entity in a metaphysical realm of existence"; and as Eddington points out, this is a problem physics itself is unable to resolve. Yet answerable or not, the question imposes itself ineluctably, seeing that what physics deals with, in the final count—what it calculates mathematically and measures by way of its empirical *modus operandi*—are in fact probabilities.

How, then, can one conceive of probabilities in realist terms? It appears that Heisenberg has put us on the right track when he observed that the Schrödinger wave function, interpreted *à la* Born as a probability wave, constitutes "a quantitative version of the old concept of '*potentia*' in Aristotelian philosophy."[22] Philosophically speaking, a probability, then, is a *potentia*: a "potency" as opposed to an "act." It is in fact a *potentia* in two senses of this Latin term: first, as something that is "potential," something that is waiting, as it were, to be actualized, and could therefore be characterized as a mere possibility; but it is also a *potentia* in the sense of a certain capacity or even power to attain the actualization to which it is thus ordered. The probability of "heads," for example, is actualized when a coin is tossed a hundred or a thousand times, and is moreover expressive of what might be termed a tendency: the tendency of fair coins to come up "heads" 50% of the time. One sees, thus, that as *potentiae*, probabilities are in fact real, or better said, *can be real.* They exist, if one may use this term, in relation to the corporeal world,[23] even as distances, or temporal durations, exist. It is crucial

---

22. *Physics and Philosophy*, op. cit., p. 41.

23. It is to be recalled that by a "corporeal" object I mean something that is to be known by way of cognitive sense perception, whereas a "physical" entity is something to be known by the *modus operandi* of physics. As the reader may be aware, the distinction between the corporeal world and the physical universe has long been fundamental to my world-view, and proves to be crucial for the interpretation

to note that we are speaking here in ontological terms, and not operationally, which is to say that the conception of probabilities as *potentiae* does not reduce to their operational definition, any more than the general concept of distance, for example, reduces to a procedure by which distances can be measured. From a philosophical point of view the concept of a real quantity precedes logically the *modus operandi* of its measurement.

One sees that in addition to its operational meaning, the mathematical formalism of physics has also an ontological significance. In fact, if the mathematical symbolism, in its totality, did not implicate an objective referent of some kind, it could not carry a pragmatic sense either; in the final count, truth and utility are *not* "here one and the same," as Bacon declared. The point is that truth comes first: truth has primacy in relation to utility, even as cause has primacy in relation to effect. The quantum description, therefore, *must* have an objective referent, even though that referent falls perforce outside the object-subject universe of physics, that is to say, transcends the physical universe itself.

However, not only the putative quantum particles—that is, their probability distributions—but other facets pertaining to the quantum formalism prove likewise to be ontologically meaningful. The single most enlightening example in that regard is doubtless what physicists term "the collapse of the state vector," something that occurs at the moment of measurement. Here is what happens. A system comprised, let us say, of a quantum particle plus an instrument of measurement, evolves, as it should, in accordance with the so-called Schrödinger equation, until the particle (to speak again in figurative terms) enters the measuring space and its presence is registered by the resultant state of the instrument. Now, at that moment—for no *physical* reason at all—the Schrödinger trajectory

of quantum theory, as I have argued in *The Quantum Enigma* (Tacoma, WA: Angelico Press/Sophia Perennis, 2012).

is violated, or as physicists like to say, is re-initialized.[24] What is the cause of this discontinuity? It derives, I say, from the fact that the instrument, by virtue of being perceptible, is perforce *corporeal*.[25] Think of it: in the act of measurement a *physical* particle is incorporated into a *corporeal* instrument! What does this entail? It implies, I maintain, that the particle is physical no more: thus incorporated, it has ceased to be a mere quantum particle and has become an actual component of a corporeal entity. As such, moreover, the putative particle has no existence apart from the instrument, which is to say, in Scholastic terms, that it participates in its substantial form.[26] Now, all this, to be sure, transcends the purview of the physicist, who continues—after the aforesaid incorporation—to view the given quantum particle as simply a quantum particle, and the instrument as simply a physical system. Yet, even so, it happens that the "transformation" of which we speak shows up on his charts: it manifests itself precisely in the aforesaid discontinuity, the so-called "collapse of the state vector." The meaning, therefore, the significance of state vector collapse, proves to be *ontological.* Simply stated, that "inexplicable discontinuity" betokens a transition from the physical to the corporeal domain. It needs, however, to be understood that what thus "corporealizes" the particle is something far removed from our customary notions; to put it in Scholastic terms: it is precisely *the act of a substantial form.*[27]

It has thus become apparent that, in its own way, physics speaks about the real world—if only one is able to listen, to understand. It would of course be absurd to suggest that quantum theory entails a complete ontology; the theory does, however, point unmistakably beyond the physical domain to the corporeal, which enters the picture by virtue of the fact that instruments of detection and mea-

---

24. The process of measurement can be described in the language of probabilities, in which case the decisive event is conceived as the incorporation, not of a particle, but of "information" in a technical sense. See Roy Frieden, *Physics from Fisher Information*, op. cit., pp. 63–111.

25. See note 23.

26. For the sake of simplicity I am suppressing the metaphysical distinction between "substances" and "mixtures," which does not change the picture.

27. I have dealt with this issue at length in *The Quantum Enigma*, op. cit.

surement are perforce perceptible. But in so doing—in the very act of "pointing beyond the physical"—quantum theory provides us with the key to the ontological understanding of the physical domain itself. The crucial fact is that the physical universe proves to be inherently *transitive*: like the probabilities to which, in the final count, it appears to reduce, it points beyond itself to something else, that is to say, to something that is *not* physical. Inherently bereft of substance, the physical *must* in fact refer to a realm in which substance is to be found. One can say that the physical as such has the nature of a sign, that in fact it is a *semantic* entity, one that is oriented, by virtue of its semanticity, towards the corporeal domain. Physics is indeed the science of measurement, as Lord Kelvin recognized long ago, and it is consequently in the act of measurement that this science reveals its nature. To be precise: *the physical as such reveals its nature in the act of measurement*. Even as an ordinary probability is realized in the toss of a coin or the roll of a die, and thus in an act which is *not* a probability, so too the physical reveals itself in a *non*-physical act.

A curious fact emerges from these reflections: having shut out from his purview, in the name of epistemic closure, the concept of the corporeal world, the physicist has, at the same time, closed the door to an understanding of the physical universe itself. We have said that physics speaks of the real world: the tragedy, however, is that the physicist, of all people, is unable to listen, unable to hear what physics itself has to tell! Reduced to its technical or "scientific" sense, physics becomes perforce "ontologically incomprehensible"; and that is the reason, after all, why one speaks of "quantum strangeness," or of "quantum paradox." It is the reason why Richard Feynman remarked that "no one understands quantum mechanics," and why Whitehead lamented that physics has turned into "a kind of mystic chant over an unintelligible universe." My point is that the very criterion of scienticity which empowers the contemporary physicist to practice his art prevents him from grasping its true significance. This is of course an unnatural condition: something has gone drastically awry. No such impasse, moreover, occurs in the case of the traditional scientist, who remains open to being, open to the mystery of cosmic existence, which is to say that whatever closure takes place on a conceptual plane in the formation of a traditional science is merely instrumental, never absolute. With the advent of modern science, on the other hand, the picture has radically changed: the concept itself has become the object, which is to say that a cognitive inversion—a veritable "*metanoia*" in reverse—has taken place, which in effect has annihilated the very possibility of authentic knowing. Only a strange kind of half-knowing is attainable under such auspices: a knowing ineluctably beset with delusion, a false knowing which alienates the knower from the real.

One sees, in conclusion, that Professor Borella's notion of "epistemic closure" proves indeed to be decisive, that in fact it provides the key to the philosophical understanding of modern science: its nature, its scope, and its implications for the human individual and for society.

4

THE ENIGMA
OF VISUAL PERCEPTION

IF VISUAL PERCEPTION proves to be singularly recalcitrant to scientific scrutiny, few subjects prove nonetheless as enlightening when the inquiry is pursued in sufficient depth and without the impediment of fallacious premises. Given that sight is the highest sense and prime means of access to the external world, all human knowledge—even the most scientific—hinges upon that cognitive act; no wonder it does not readily submit to scrutiny! To be sure, the scientific investigation of visual perception was well under way by the time Hermann von Helmholtz brought out his celebrated multi-volume *Handbuch der physiologischen Optik* between 1855 and 1866; and it is needless to add that during the following century and a half the scientific literature pertaining to this domain has grown exponentially. New and formidable disciplines, moreover, have entered the field, notably neurophysiology, computer science, and the theory of artificial intelligence; there is, however, reason to question whether the application of even these sophisticated means has brought us any closer to an understanding of the perceptual act: of how in fact we "*see.*"

It is not my intention in the present chapter to delve into the history of the cognitive sciences pertaining to visual perception; my object, rather, is to report and comment upon a radical paradigm shift proposed during the latter half of the twentieth century. What aroused my interest in this new approach to perception is not only the solidity of its empirical basis, but the fact that the resultant theory turns out to be incurably non-bifurcationist and consequently inimical to the Cartesian worldview. As some readers may surmise,

69

I am referring to the so-called "ecological" theory of visual perception propounded by the late James J. Gibson, a Cornell University psychologist who devoted half a century to the study of that subject. It would however be misleading to characterize his theory as a "breakthrough in the psychology of perception"; it is rather to be viewed as a brand new start, which from the outset rejects as chimerical the central premise upon which the various "non-ecological" approaches are based.

It may be well to recall, by way of introduction, that Gibson began to formulate his theory during the 1940's while engaged in research having to do with the design of tests that would ascertain the ability of a prospective pilot to fly a plane, and in particular, to land it visually without crashing. It thus became necessary to understand how one perceives certain parameters, such as the "aiming point" of a landing approach. Inasmuch as visual perception derives, according to the conventional wisdom, from the so-called retinal image, it was natural to define the aiming point in terms of retinal motion and gradients of retinal velocity; it turns out, however, that this cannot be done: "Such a statement," Gibson informs us, "cannot be made exact and leads to contradiction." (182)[1] It is findings such as this that eventually led Gibson to abandon the postulate that visual perception is image-based; and so began his search for its actual basis, whatever that might turn out to be. In time Gibson concluded that perception derives from a hitherto unrecognized structure inherent in the ambient light: and this is the discovery that inaugurates his so-called "*ecological*" approach to visual perception. On this new-found basis Gibson was able to resolve numerous problems that had hitherto proved recalcitrant, beginning with the conundrum of the aiming point: "It turns out that the aiming point of any locomotion is the center of the centrifugal flow of the ambient optic array. Whatever object or spot on the ground is specified at that null point is the object or spot you are approaching. This is an exact statement." (182) It emerges that the information which enables us to perceive objects, events and

---

1. Page references in parentheses refer to Gibson's major work, *The Ecological Theory of Visual Perception* (Hillsdale, NJ: Lawrence Erlbaum Publishers, 1986).

motions is given, not in a stipulated visual image—be it retinal, cerebral or mental—but *objectively* in what Gibson terms the "ambient optic array": it is thus to be found, not inside the head, but outside, in the external world. This, in brief, is the fateful discovery to which Gibson was led early in his career by way of the mundane business of screening prospective pilots; it can be said, in retrospect, that a radically new understanding of visual perception lay concealed in these unpretentious beginnings.

Inevitably, along with the notion of the "visual image," many other basic teachings of cognitive psychology prove to be likewise untenable; for example, the conventional theory of depth perception. It had been supposed that the perception of depth results from the imposition of a third dimension upon a flat visual field, a task which supposedly was accomplished through the utilization of so-called "cues." It now turns out, however, that the ambient optic array itself specifies the surfaces, textures and layout of the environment, which is to say that the third dimension is not in fact constructed or somehow deduced from a flat image, but is directly perceived: depth perception, one finds, is not actually a two-stage process, as visual-image psychology was forced to assume. But this recognition entails another, the most surprising of all: since the retinal image can give rise to at most a 2-dimensional view, one is led to conclude that visual perception is not actually based upon retinal stimulation. The fact is that a single-stage theory of depth perception negates the very basis of the conventional scientific approach to visual perception! What stimulation of receptor surfaces engenders are *sensations*, properly so called; the point, however, is that *perception is not sensation-based*. Sensations do of course have a role to play in the integral process of perceiving; yet they are not what is directly perceived.

This revolutionary discovery absolves the cognitive scientist at one stroke from the very problem he had labored the hardest to resolve, which is to understand how percepts are produced out of sensations. Until now that daunting task had imposed itself ineluctably, which is to say that sensation-based theories of perception are perforce constructivist: one evidently requires a process of some kind to supply in the percept all that is missing in a visual image, beginning with the dimension of depth. As Gibson explains: "They postulate activities to supplement sensations, or to correct them, or to interpret them, or to organize them, or to fuse them with memories, or combine them with concepts, or impose logic on them, or construct a model of the world from them (the list could go on and on)."[2] The journals of cognitive science are filled to the brim with the fruits of these prodigious labors; yet, from Gibson's point of vantage, one is attempting to solve a problem which in reality does not exist. What is perceived, according to that theory, are not constructs or representations superimposed supposedly upon a visual image, but quite simply external objects and events specified in the ambient optic array. It behooves us now to take a closer look at that claim.

It is in *The Ecological Approach to Visual Perception*, first published in 1978, that Gibson expounds his doctrine in its definitive form. He begins with the decisive observation that "Physics, optics, anatomy and physiology describe facts, but not facts on the level appropriate for the study of perception." (xiii) Rejecting the prevailing modes of reductionism, he arrives at a recognition of "levels," an idea consonant with the traditional conception of ontological hierarchy. It appears, however, that Gibson attained to a hierarchic view of reality, not on metaphysical grounds, but was led, as a hard-headed empiricist, to recognize that unfounded premises had long been leading the cognitive sciences astray. Like Einstein, Heisenberg, and other pioneers of twentieth-century science, he felt that his discipline needed to be purged of baseless and counterproductive hypotheses, and it was in the spirit of "back to the facts" that Gibson finally reached his startling conclusions.

The first thing that needed to be done was to forge suitable conceptions concerning the perceived world "on a level appropriate for the study of perception." The new science-to-be required a technical

---

2. "The Myth of Passive Reception: A Reply to Richards," *Philosophy and Phenomenological Research* 37 (1976), p. 234.

jargon of its own, a terminology divested of the prevailing reductionist connotations. As his primary notion, Gibson selected the term "environment," which he took to signify the *perceptible* world; and from the start he recognized that "this is not the world of physics." (2) The question arises, of course, how "the world of physics" is related to the environment, "the world we perceive"; and as might be expected, this is an issue which Gibson does not pursue in depth. He does touch upon it occasionally; for instance when he observes:

> All sorts of instruments have been devised for mediating apprehension. Some optical instruments merely enhance the information that vision is ready to pick up, others—the spectroscope, for example—require some inference; still others, like the Wilson cloud chamber, demand a complex chain of inference.... Indirect knowledge of the metric dimensions of the world is a far extreme from direct perception of the affordance dimensions of the environment. Nevertheless they are both cut from the same cloth. (260)

Whatever it might mean, in this context, to be "cut from the same cloth," by distinguishing between the "metric dimensions" of interest to physics and the "affordance dimension" of ecological theory, Gibson wishes to affirm a non-reductionist view of the environment. It is clear that from the outset he rejects the Cartesian postulate of bifurcation: what we perceive—what we actually apprehend in visual perception—is not inside, but outside the head, a notion which, in its unabashed realism, shocked just about everyone and scandalized most of his peers. And yet, what could be more empirical, more truly scientific? I will note, in passing, that Gibson's conception of the environment is tantamount to what I refer to as the corporeal world, which is to say that our respective doctrines are in fact compatible.

With his definition of "environment" in place, Gibson proceeds to specify the primary divisions of that domain in suitable "ecological" terms; and I find it significant that, in so doing, he reverts to the so-called "elements" of traditional cosmology: "Let us begin," he writes, "by noting that our planet consists mainly of earth, water and air—a solid, a liquid, and a gas." (16) Based upon these primary distinctions, Gibson goes on to define the key notions of his theory; a *surface*, for example, is an interface between two elements, beginning with the interface between the Earth itself and air, which Gibson terms "the ground." And one might add that although he does not mention the fourth traditional element (that is, "fire") by name, this too enters the picture in a most essential way: for we may take it that the latter refers to radiation, and therefore to light, which is of course precisely the element that enables visual perception to take place.

Following upon the definition of "surface" and "ground," Gibson goes on to define the ecological concept of *"substance,"* which he specifies as "matter in the solid or semisolid state." As is to be expected, substances are characterized ecologically by such properties as hardness, viscosity, cohesiveness, density, plasticity and the like, all of which however pertain to surfaces: "The surface," says Gibson, "is where most of the action is." (23) In addition to substances and surfaces, there are also *media*, which are relatively insubstantial, and are characterized by the fact that they afford locomotion. For man, and for terrestrial animals, air constitutes the only medium, which is to say that water is classified as a substance. It is to be observed that in Gibson's ecological theory, the medium takes the place of space, and is endowed with a vertical axis defined by the pull of gravity, plus an east-west axis[3] specified by the rising and setting of the Sun: "This fact," Gibson points out, "reveals another difference between medium and space, for in space the three reference axes are arbitrary and can be chosen at will." (8) Along with an absolute frame of reference, ecological theory recognizes rest and motion as being likewise absolute: "The environment is simply that with respect to which either locomotion or a state of rest occurs, and the problem of relativity does not arise." (75) One is

---

3. It is may not be without interest to note that Gibson's ecological notion of "the medium" corresponds to René Guénon's conception of "qualified space": "It is the notion of direction which without doubt represents the real qualitative element inherent in the very nature of space, just as the notion of size represents the quantitative element; and so space that is not homogeneous, but determined and differentiated by its directions, may be called 'qualified' space." *The Reign of Quantity and the Signs of the Times* (San Rafael, CA: Sophia Perennis, 2004), pp. 34–35.

beginning to see that the ecological universe proves, in many ways, to be in fact *Aristotelian.*

The information upon which visual perception is based, Gibson maintains, is given in the ambient light; to recognize and investigate the relevant structures, however, one requires an *ecological optics* which differs significantly from the physical. The difference stems, first of all, from the fact that ecological optics is concerned with "ambient" as distinguished from radiant light: "Radiant light diverges from an energy source; ambient light converges to a point of observation.... Radiant light is energy; ambient light can be information...." (51) It *can* be precisely because it is conditioned by the environment: "Only because ambient light is structured by the substantial environment can it contain information about it." (86) What is needed for visual perception are structures pertaining to ambient light that specify portions or aspects of the substantial environment; and these are what ecological optics is about. What stands at issue is a nested sequence of solid angles with apex at the point of observation; and as Gibson informs us, the idea goes back to the "visual cones" of Euclid and the "pyramids" of Ptolemy, upon which whatever science of visual perception the Greeks may have possessed was apparently based. Eventually the optics of visual solid angles came however to be replaced by a geometrical optics based upon rays, which can indeed explain such things as the operation of a camera, but proves to be unsuited for the study of how we perceive. It turns out that an optics based upon point-to-point correspondences does not pertain to "a level appropriate for the study of perception," but constitutes what Gibson terms a *physical* as distinguished from an *ecological* optics. With the ascendance of the former in modern times the structures in ambient light permissive of visual perception fell consequently into oblivion, and remained unsurmised till the latter half of the twentieth century, when Gibson's discovery of "ecological optics" brought them back into view.

It needs to be clearly understood that the camera paradigm does not take us far in the study of perception; as Gibson explains:

> The information for perception of an object is not in its image. The information in light to specify something does not have to

resemble it, or copy it, or be a simulacrum or even an exact projection. Nothing is copied in the light to the eye of the observer, not the shape of the thing, not the surface of it, not its substance, not its color, and certainly not its motion.[4] But all these things are specified in the light. (304)

The full gamut of information required for visual perception of the environment is given in a hierarchy of nested solid angles at the point of observation, or more precisely, in the field corresponding to possible points of observation. As a rule, ambient light carries a structure of this kind rich enough to specify the relevant portions of the substantial environment. There are of course exceptional cases, as in a dense fog, for example, or a space in which all incoming light has been filtered through some translucent substance such as "milk glass," the effect of which is precisely to eliminate the structures in question through randomization of the light. Under normal conditions, however, ambient light is structured by the substantial environment so as to specify the very features of that environment by which it is thus formed.

Theories of visual perception are subject to empirical verification by way of psychophysical experiments. Typically a subject is exposed to visual stimuli designed to simulate the factors thought to be responsible for the perception of certain parameters; as Gibson observes: "In order to study a kind of perception an experimenter must devise an apparatus that will display the information for that kind of perception." (170) Different theories of visual perception, however, stipulate different kinds of pertinent information, a fact which in principle renders such theories testable. How, for example, does one perceive the size of a distant object? According to sensation-based theory, the size of the object must be somehow deduced from the primary data given in the retinal image, an assumption

4. Gibson's point is that optical and mechanical "motion" are entirely different: "The two kinds of 'motion', physical and optical, have nothing in common and probably should not even have the same term applied to them." (103)

which leads quite naturally to the conclusion that the perceptions of size and distance are based upon the laws of linear perspective familiar to artists since the Renaissance. What interests Gibson, on the other hand, are not the shapes and sizes of patches given in a retinal image, but the relations of external objects to each other and to their common ground. Here, then, is one of the early experiments he carried out to test his theory: In a large plowed field without furrows, receding almost to the horizon, he planted a stake at distances up to half a mile, and asked observers to judge its size. It is to be noted that linear perspective has been essentially ruled out by the conditions of this experiment; yet the perceived size of the stake did not decrease with distance, even when the stake was about a third of a mile distant and was becoming hard to see: "The judgments became more variable with distance but not smaller. Size constancy did not break down. The size of the object only became less definite, not smaller." (160)

But although these findings are at variance with sensation-based theory, it appears that Gibson did not consider them definitive: it was not his nature to draw conclusions on the basis of a single experiment. Eventually, however, in light of experimental evidence "accumulated in the last twenty-five years," he returned to the aforesaid experiment to observe:

> The implication of this result, I now believe, is that certain invariant ratios were picked up unawares by the observers and that the size of the retinal image went unnoticed. No matter how far away the object was, it intercepted or occluded the same number of texture elements on the ground. This is an invariant ratio. (160)

The perception of size, in this instance, was apparently accomplished by way of a hitherto unrecognized invariant given directly in the ambient optic array.

This brings us to the heart of Gibson's theory: the idea, namely, that perception results from the pickup of *invariants* given in ambient light. Up till then it had been assumed that perception is based upon shapes (what cognitive psychologists term "forms") first given in the retinal image, an assumption which leads, as we have noted before, to a two-stage view of perception. For decades researchers had investigated what was termed "form discrimination" by means of psychophysical experimentation; "My objection to this research," Gibson writes, "is that it tells us nothing about perceiving the environment." (150) What he means, in effect, is that the research in question relates to the visual interpretation of *pictures*, of two-dimensional pictorial displays; and to be sure, as such these studies do provide correct and potentially useful information. The environment, however, is something very different from a pictorial display; and therefore, if we do perceive the environment (as Gibson claims), the optical information upon which that perception is based must differ fundamentally from the "optical cues" studied by visual-image psychologists. The latter point to their success in the investigation of "form discrimination" as a vindication of their theory, forgetting that "this tells us nothing about perceiving the environment."

It is evident that visual-image psychology employs physical optics based upon rays; as Gibson goes on to point out:

> This theory of point-to-point correspondence between an object and its image lends itself to mathematical analysis. It can be abstracted to the concepts of projective geometry and can be applied with great success to the design of cameras and projectors, that is, to the making of pictures with light. The theory permits lenses to be made with smaller "aberrations," that is, with finer points in the point-to-point correspondence. It works beautifully, in short, for the images that fall on screens or surfaces and are intended to be looked at. But this success makes it tempting to believe that the image on the retina falls on a kind of screen and is itself something intended to be looked at, that is, a picture. It leads to one of the most seductive fallacies in the history of psychology—that the retinal image is something to be seen. (59–60)

Taking the image-paradigm at its face value, one needs in effect to postulate a "little man" inside the head who looks at the stipulated image, a notion which leads in principle to an infinite regress, that is to say, to an indefinite sequence of "little men," one inside the preceding. On the other hand, if one adopts a more sophisticated approach based upon neurophysiology, one arrives at a correspondence between points of stimulation on the retina and what Gibson terms "spots of sensation in the brain," spots which are characterized by brightness and color alone. "If so," Gibson goes on, "the brain is faced with the tremendous task of constructing a phenomenal environment out of spots differing in brightness and color. If these are what is seen directly, what is given for perception, if these are the data of sense, then the fact of perception is almost miraculous." (61) By way of contrast, Gibson goes on to enunciate his own position:

> It is not necessary to assume that *anything whatever* is transmitted along the optic nerve in the activity of perception. We need not believe that either an inverted picture or a set of messages is delivered to the brain. We can think of vision as a perceptual system, the brain being simply a part of that system. The eye is also a part of the system, since retinal inputs lead to ocular adjustments and then to altered retinal inputs, and so on. The process is circular, not a one-way transmission. The eye-head-brain-body system registers the invariants in the structure of ambient light. The eye is not a camera that forms and delivers an image, nor is the retina simply a keyboard that can be struck by fingers of light.

It is to be noted that the shift from retinal receptors and afferent nerve bundles to the "eye-head-brain-body" complex conceived as a single perceptual system parallels, on the side of the perceiving organism, the transition from the physical world to the environment: it turns out that the new concept of perceptual system corresponds indeed to a level "appropriate for the study of perception." What stands at issue, once again, is the repudiation of a reductionism based upon the breaking up of a whole and its subsequent demotion to the sum of the resultant parts. As always, what is lost in the bargain is precisely the *substantial form* of the whole in question, a matter to which we shall return in the sequel. For the moment it suffices to observe that the shift from receptors and afferent nerve bundles to the newly-conceived perceptual system obviates the need for a "little man," and absolves us also from having to explain how one constructs "a phenomenal environment out of spots differing in brightness and color": the crucial point is that *one is now confronted, not by an assemblage of neurons, each in its own state, but by a perceptual system which does not in fact reduce to the sum of its parts.*

The perceptual system is designed for the pickup of information given in ambient light, and specifically for the apprehension of invariants, that is to say, of structural elements in the ambient optic array which persist in time and remain unaffected by changes in visual perspective. But this implies that time, or better said, *motion* enters the picture in an essential way; nothing, in fact, can be perceived "in an instant." As Gibson points out:

> The eye is never literally fixed. It undergoes a series of miniature movements or microsaccades.... Looking is always exploring, even so-called fixation.... The visual system hunts for comprehension and clarity. It does not rest until the invariants are extracted. (212, 220)

It is indeed movement that discloses the invariants, the things that we actually perceive. Even true colors, Gibson maintains, constitute invariants which emerge as lighting changes, just as the true shape of a surface emerges as the perspective shifts. (89)

In addition to objects and their qualities we also perceive events. The latter are specified, not by invariants, but by a disturbance in the invariant structure, precisely. Yet even so, events are likewise specified in the ambient optic array, and are perceived *directly*. It is not a matter of unifying a sequence of instantaneous perceptions, as sensation-based theories had assumed, but a question, once again, of information pickup. "Perceiving," writes Gibson, "is a registering of certain definite dimensions of invariance in the stimulus flux, together with definite parameters of disturbance. The invariants specify the persistence of the environment and of oneself. The disturbances specify the changes in the environment and of oneself."

(249) I will note in passing that the words "and of oneself" are highly significant, inasmuch as they imply that exteroception and proprioception are complementary and thus inseparable functions. What I wish especially to draw attention to, however, is the fact that there exists, according to Gibsonian theory, a direct perception of persistence, which is something entirely different from the persistence of a perception. Given the far-reaching significance of this fact, it will be well to describe a pivotal experiment supportive of this contention.

The experiment, first performed by G.A. Kaplan in 1969,[5] involves a cinematographic display in which a photograph of a textured surface was altered, frame by frame, so as to produce a dissimilar texture on one side of a moving invisible line. The display was intended to simulate the optical information specifying the progressive occlusion of one surface by another. All observers perceived one surface going *behind* another, or coming from behind another when the process was reversed. "In short, one surface was seen in a legitimate sense behind another at an occluding edge." (190) When the film was stopped the edge perception ceased and was replaced by the perception of one continuous yet divided surface. Now, whatever this result may have suggested to other investigators, Gibson recognized therein a disproof of the classical theory: "For we are not allowed to say that a hidden surface is *perceived*; we can only say that it is remembered.... If an occluded surface is perceived, the doctrine is upset." (189) Gibson insists that an occluded surface *can* be perceived, that there can in fact be perceptions without corresponding sensations.[6] It is this remarkable claim that we need now to reflect upon. The key to the problem, as we shall see, lies in the recognition that what we perceive are not images, but *invariants*. The following explanations—which richly deserve to be quoted at length—may help to make this clear:

---

5. "Kinetic Disruption of Optical Texture: The Perception of Depth at an Edge," *Perception and Psychophysics* 6, pp. 193–98.

6. Discoveries in the field of neurophysiology (e.g., "subjective contours") have since verified Gibson's conclusion. See Chapter 5, pp. 112–114.

The old approach to perception took the central problem to be how one could see into the distance and never asked how one could see into the past and future. These were not problems for perception. The past was remembered, and the future was imagined. Perception was of the present. But this theory has never worked.... The environment seen-at-this moment does not constitute the environment that is seen. Neither does the environment seen-from-this-point constitute the environment that is seen.... It is obvious that a motionless observer can see the world from a single fixed point of observation and can thus notice the perspective of things. It is not so obvious but it is true that an observer who is moving about sees the world at *no* point of observation and thus, strictly speaking, *cannot* notice the perspective of things. The implications are radical.... The world is not viewed in perspective. The underlying invariant structure has emerged from the changing perspective structure.... It sounds very strange to say that one can perceive an object or a whole habitat at no fixed point of observation, for it contradicts the picture theory of perception and the retinal image doctrine on which it is based.... But the notion of ambulatory vision is not more difficult, surely, than the notion of successive snapshots of the flowing optic array taken by the eye and shown in the dark projection room of the skull. (195, 197)

One sees from these remarkable observations that, in Gibson's theory, invariants replace the visual image as the founding conception.

It is to be noted that what may be termed an ecological conception of time is implicit in the Gibsonian theory, a fact of which Gibson himself was keenly aware. What constitutes time, ecologically speaking, are events: "Events are perceived, time is not." (101) Elsewhere Gibson points out that "The stream of experience does not consist of an instantaneous present and a linear past receding into the distance ... there is no dividing line between the present and the past, between perceiving and remembering." (253) And let us note: only on this basis could there be a *direct* perception of persistence and events, as Gibson claims. It seems that time, too, needs to be conceived "on the level appropriate to the study of perception." Even as substances must not be reduced to so-called atoms, so too the flux of time, it appears, must not be "atomized" in the manner of physics, that is to say, reduced to "instants" the way a line is reduced to points.[7]

A few words, now, concerning the Gibsonian notion of "affordances." On the face of it the idea is unremarkable: an affordance is simply something pertaining to the environment that is afforded or "offered" to a percipient. The medium, for instance, affords locomotion, an enclosure affords shelter, and a fruit affords eating. What however renders that seemingly innocuous notion difficult and indeed profound is the fact that affordances defy the customary subject-object dichotomy: they are not exclusively objective, because they relate intrinsically to a subject, but neither are they merely subjective, because they derive from the object. It is on account of this dual nature that Rom Harré has applied the concept to the interpretation of quantum theory: "What a system affords," Harré points out, "is relative to the nature of the being which interacts with it, in particular what states it is capable of taking up. Affordances are dispositions of physical things relativized to that with which they interact."[8] It appears that, in the final count, what quantum physics deals with are precisely *affordances*: "As a fundamental or near fundamental physical theory," Harré goes on to say, "quantum field theory must deal in affordances." And it turns out— as indeed one might expect—that this fundamental recognition sheds light on the conundrums of quantum mechanics, beginning with Bohr's complementarity principle and the idea of "virtual particles," which now become philosophically comprehensible.

---

7. It is not without interest to note that here too, in the prevailing conception of time, we encounter the mark of René Descartes, for it is by way of his "analytical geometry" that the continuum came to be conceived as an infinite set of points, which is to say that in earlier times it was not thus dismembered.

8. *Philosophical Foundations of Quantum Field Theory* (Oxford: Clarendon Press, 1990), p.67.

It is interesting that Gibson came upon his now famous concept by way of gestalt psychology: "The gestalt psychologists," he tells us, "recognized that the meaning or the value of a thing seems to be perceived just as immediately as its color." (138) As Koffka himself has put it: "Each thing says what it is ... a fruit says 'eat me'; water says 'drink me'; thunder says 'fear me'; and a woman says 'love me.'"[9] Psychologists have coined various terms to describe this "something" in objects which issues invitations—such as Kurt Levin's inimitable "*Aufforderungscharakter*"—without however managing to break the subject-object dichotomy; as Gibson goes on to explain:

> The accepted theories of perception, to which the gestalt theorists were objecting, implied that no experiences were direct except sensations and that sensations mediated all other kinds of experience. Bare sensations had to be clothed with meaning. The seeming directness of meaningful perception was therefore an embarrassment to the orthodox theories, and the Gestaltists did right to emphasize it. They began to undermine the sensation-based theories. But their own explanations of why it is that a fruit says "Eat me" and a woman says "Love me" are strained. The gestalt psychologists objected to the accepted theories of perception, but they never managed to go beyond them. (140)

The point is that Gibson himself did manage to "go beyond" the accepted theories of perception, and did so precisely through the clear-minded recognition that "The object offers what it does because it is what it is." (139) The affordance, therefore, belongs as much to the object as to the subject: it is not simply "phenomenal" in the contemporary sense, as it had been for the Gestaltists. As Gibson points out: "For Koffka it was the *phenomenal* postbox that invited letter mailing, not the *physical* postbox. But this duality is pernicious." There is only in fact *one* postbox, and "Everyone above the age of six knows what it is for and where the nearest one is."

This matter having been settled, Gibson can say, quite simply, that "Affordances are properties taken with reference to the observer. They are neither physical nor phenomenal." (143) They

---

9. Kurt Koffka, *Principles of Gestalt Psychology* (NY: Harcourt Brace, 1935), p. 7.

are in fact *ecological*, and as such they can be perceived:

> The perceiving of an affordance is not a process of perceiving a value-free physical object to which meaning is somehow added in a way that no one has been able to agree upon; it is a process of perceiving a value-rich ecological object.... The central question for the theory of affordances is not whether they exist, but whether information is available in ambient light for perceiving them. (140)

And as the reader will surely have come to expect by now, it turns out that the requisite information *does* exist. There can be no doubt, moreover, that Gibson is amply justified in referring to this discovery as "the culmination of ecological optics."

As in twentieth-century physics, so in the Gibsonian theory the idea of "information" has emerged as a pivotal conception. But here again one finds that the "ecological" concept differs fundamentally from the physical: given that the ecological information in the optic array conveys all the qualities and affordances of the visually perceptible world, the ecological must differ from the physical by virtue of the fact that it evidently does not reduce to the quantitative order.

According to Gibson's theory, what we perceive is actually the environment. Sensation-based theories, on the other hand, cannot be objective: what a sensation specifies, after all, is not an external reality, but the state of a receptor, which is something else entirely. And as Gibson points out, it is precisely because sensation does *not* convey knowledge of the external world that sensation-based theories are perforce constructivist. But the end-result of "processing" can at best be a *representation* of some kind. If the terminus of visual perception is indeed external, as Gibson insists, then it follows that sensation-based theories are *ipso facto* false: for needless to say, no amount of construction or processing can give rise to an object or event pertaining to the environment. It emerges that in rejecting the long-standing axiom that perception is sensation-based, and replacing it with his revolutionary notion of "information pickup,"

Gibson has at last opened the way to a *realist* theory of visual perception.

Certainly he admits that sensations exist, and are caused by the stimulation of receptors; what he denies is simply that perception is sensation-based, to say it again. It needs thus to be recognized that there are different modes of visual awareness. There is, first of all, the direct or immediate kind; and within this category there are objective and subjective modes. As Gibson explains:

> There can be direct or immediate awareness of objects and events when perceptual systems resonate so as to pick up information, and there can be a kind of direct or immediate awareness of the physiological states of our sense organs when the sensory nerves as such are excited. But these two kinds of experience should not be confused, for they are at opposite poles, objective and subjective. There can be an awareness of other bodily organs than sense organs, as in hunger and pain, and these are also properly called sensations.[10]

In addition to direct modes of visual perception, moreover, there are also indirect or mediated modes, and with reference to these Gibson does not rule out the possible relevance of a constructivist approach; what he categorically rejects are simply constructivist theories of perception: "The fallacy," he tells us, "is to assume that because sensory inputs convey no knowledge, they can somehow be made to yield knowledge by 'processing' them." (253) There is no question that mediated modes of visual awareness, such as reminiscence, expectation, imagination, fantasy, and dreaming do occur; what Gibson denies is that they are instances of perceiving: "They are kinds of visual awareness other than perceptual." (254)

Gibson's astonishing claim is that our normal belief is correct: what we actually perceive is not an image, not a representation of some

10. *Reasons for Realism: Selected Essays of James J. Gibson*, edited by R. Reed and R. Jones (Hillsdale, NJ: Lawrence Erlbaum Publishers, 1982), p. 380.

kind, not a thing that exists in the brain or mind of the percipient, but is indeed an external object or event. Now, this is of course a *philosophic* claim; and yet Gibson propounds it on scientific grounds. Here is one of his arguments: "An object can be scrutinized.... No image can be scrutinized—not an afterimage, not a so-called eidetic image, not the image in a dream, and not even a hallucination." (257) What Gibson has in mind when he speaks of "scrutiny" with reference to an object is above all the fact that, by way of perception, we can "tap into" the unlimited store of information given in the ambient optic array. Obviously we can be deceived, as in the case of cinematographic display; but the illusion vanishes the moment we begin to explore the optic array in a region of the ambient space. Since it is mathematically impossible to simulate *all* perceptually relevant structures given in ambient light, it follows that the illusion arising from pictorial display of whatever kind cannot bear the scrutiny which in fact takes place in normal perception. Similar remarks apply to other kinds of illusory experiences. Think of the man who mistook the proverbial rope for a snake: clearly, it is the rope, and not the snake, that bears scrutiny. The snake, in this instance, is not "given" in the ambient optic array, but is evidently superimposed: it is not in fact perceived, but imagined, one can say, and pertains in any case to a fundamentally different kind of visual awareness.

"I suggest," writes Gibson, "that perfectly reliable and automatic tests for reality are involved in the working of the perceptual system." (256) It is to be noted that the term "automatic" carries weight: it is not a question of reasoning, or of a conscious interrogation. Admittedly the pickup of information constitutes an intelligent act; but it is not ratiocinative: young children and animals as well are capable of perception. So long as a perceptual system is unimpaired and unhindered, it is bound, under normal conditions of illumination, to arrive at objectively valid perceptions. What is required for information pickup is a search, a certain scanning of the optic array, which is precisely what a perceptual system is in fact designed to do; and let us note, once again, that the concept of "movement" enters the picture in an essential way. It is not the function of a visual perceptual system to produce snapshots like a camera; it is designed, rather, to move, to search, to scan: only thus can it detect *invariants*, which is precisely what, according to Gibson's theory, perception is about. It follows, as he points out, that "a criterion for *real* versus *imaginary* is what happens when you turn and move." (257)

One sees that the Gibsonian theory presents itself as a rediscovery of realism, and indeed, of "naïve realism," as one might say.[11] And this raises an intriguing question: If a scientifically sound theory of visual perception proves thus to be supportive of realism, might not the demise of realism in Western philosophy, beginning with Descartes, be the result of a scientifically spurious concept of visual perception: a theory, namely, based upon the camera paradigm? If visual perception does in fact constitute our basic means of access to the external world, it stands to reason that a paradigm that locates percepts "inside the head" does evidently favor non-realist modes of philosophy, be they Cartesian, idealist or skeptical.

Gibson's realism, it appears, is empirically based. What sets him apart is his passion for dealing directly with the facts, and his willingness to jettison assumptions of long standing when they fail to pass empirical muster. His approach to cognitive psychology is therefore comparable in a way to the quantum revolution in physics, which likewise originated in a "return to the facts of observation." The young Heisenberg was presumably the first to recognize that the Newtonian and post-Newtonian world of particles and fields is not in fact what we observe, not what actually confronts us on an experimental plane; and so he set about to forge concepts which do accord with the experimental facts. The same holds true, I say, in the case of Gibson's "ecological" theory of visual perception: it too is based upon conceptions "forged in the crucible of empirical facts," ideas that prove to be perhaps even more inimical to the scientific *status quo*. Thus it was by way of research relating to such

---

11. Gibson has dealt with this issue explicitly in a number of articles; see *Reasons for Realism: Selected Essays of James J. Gibson*, op. cit.

concrete issues as the perception of an aiming point that he came to recognize the chimerical nature of sensation-based theories and arrived at the startling realization that what we perceive is not a "processed" plane image, but is in fact a three-dimensional environment. As Heisenberg had found that there *are* no classical particles, so Gibson came to realize that there is actually no visual image in perception. Cognitive scientists, it turns out, had accepted that notion uncritically, and have labored ever since to extricate themselves from the resultant quandary. To be sure, visual-image psychology has had its triumphs, its domain of success, which has to do with such things as the perception of pictorial displays and the design of spectacles; and whereas these accomplishments can hardly be compared with the immense achievements of pre-quantum physics, the fact remains that they likewise have served to bestow an aura of scientific legitimacy upon the theories in question. Like the young Heisenberg, Gibson too was obliged to confront a *status quo* buttressed by seemingly compelling evidence. I find it remarkable, moreover, that both were obliged in the end—each in his own way—to abandon the norm of causal explanation, an incredible step for a scientist! Quantum theory, as we know, has turned "acausal" when it comes to such things as the deflection of an electron passing through a slit; it insists, in fact, that there *can be* no mechanism that accounts for the phenomena in question. What renders the Gibsonian theory "acausal," on the other hand, is the fact that pickup of information—to which visual perception is finally reduced—cannot be accounted for on the level of neurophysiology (which evidently constitutes the only basis upon which a physical causality might in this instance be conceived).

Perhaps the most striking parallel, however, between the respective contributions of Heisenberg and Gibson derives from the fact that quantum indeterminacy, when viewed in light of David Bohm's approach, is likewise associated with a pickup of information: the pickup, namely, of what Bohm terms "active information," said to be accomplished by a mysterious "pilot wave." It is true that by means of this conception Bohm was able to reinstate a strict causality, but only on a formal level which is not, properly speaking, empirical. It can therefore be said that both quantum mechanics and the Gibsonian theory of perception entail a pickup of information that defies explanation in causal terms. The two "revolutions" may in fact be seen as complementary aspects of a single decisive event: the intrusion, namely, of *information* as an essential and indeed irreducible element in our scientific understanding of reality.

This brings us to a curious fact that may be worth mentioning: like most scientists of the twentieth century, Gibson was a convinced Darwinist. What I find surprising, on the other hand, is that his Darwinist convictions have apparently proved beneficial in his quest of truth: it seems that his fundamental distinction between the environment and the physical world was motivated, not by ontological concerns, but by Darwinist assumptions. As Gibson himself explains:

> When one studies the evolution of the "senses" in animals a puzzle appears in that they seem to have evolved not to yield sensations, but perceptions. For example, there is no survival value in being able to distinguish one wavelength from another (pure color), but there is great value in being able to distinguish one pigmented surface from another in variable illumination. In short, the survival value of the "senses" is found in the ability of animals to register objects, places, events and other animals: that is, to perceive.[12]

Certainly the force of Gibson's observation does not hinge upon Darwinist premises: "the ability of animals to register objects, places, events and other animals" is obviously essential regardless of whether evolution has occurred, which is to say that there could be no animal life based upon mere sensations, nor in a world deprived of its "ecological dimensions." Yet even if Darwinism may have set him upon the path of discovery, it is finally incompatible with Gibson's theory; for whereas the latter hinges, as we have seen, on an information-theoretic understanding of perception, it is to be noted that the concept of "information" proves ultimately fatal to

---

12. "The Survival Value of Sensory Systems," *Biological Prototypes and Synthetic Systems* 1 (1962), pp. 230–33.

Darwinism.[13] More directly to the point, however, is the fact that by his adherence to the Darwinist postulate Gibson has unwittingly closed the door to a *metaphysical* understanding of his own theory: the kind we shall now attempt to delineate.

Viewed from a traditional point of vantage, the notion of "information" that has emerged in various branches of contemporary science as a basic conception can be recognized as a rediscovery of what the Scholastics termed *"forms."* This is not to say that modern science has now arrived at an understanding of *morphe* in the Aristotelian sense, or even that it *can* do so; my point, rather, is that "information" in the scientific sense does ultimately refer to forms, whether scientists are cognizant of the fact or not. One way to discern the connection is to note that the idea of information entails the concept of a *non-physical* transmission: something is indeed conveyed, but without any movement through space. This holds true, moreover, even when the information in question is conceived as something transmitted over a channel of communication, as in the case of Shannon theory: for even then it turns out that another transmission takes place, one that has nothing to do with any channel of communication.[14] Consider the example of a verbal message: the spoken or written words may indeed be transmitted over a channel, yet what counts in the end is what the message affirms, what it signifies. *The essence of information resides in its capacity to signify*; and that is where "forms" (in the Scholastic sense) enter the picture.

13. As William Dembski has rigorously proved (and the Darwinist establishment has so far adamantly refused to acknowledge), the presence of "complex specified information" in the genome of even the simplest organism cannot be explained in terms of the Darwinist mechanism.

14. In David Bohm's version of quantum theory that "non-physical" transmission is supposedly accomplished by the stipulated pilot wave; the problem, however, is that this so-called wave (postulated precisely for the stated purpose) eludes detection. It does so, moreover, not because no one has yet been clever enough to detect it, but because it is in principle undetectable. This "principial indetectability" is tantamount to a scientific recognition that the transmission in question is indeed "non-physical." In truth no movement through space is involved.

"The most incomprehensible thing about the universe," Albert Einstein once observed, "is that it is comprehensible." Yes, the great physicist did have cause to wonder; for if indeed the universe were made up simply of particles and fields—as apparently he imagined— it could not in fact be perceived, nor would it be comprehensible to the scientist or to anyone else: what renders the world perceptible to sentient beings and intelligible to man is the presence therein of *forms*, precisely. It is moreover to be understood that these forms are not subject to the bounds of space and time, and must consequently be distinguished categorically not only from the physicist's particles and fields, but from corporeal entities as well: in a word, they are not "things." Yet *it is forms that constitute things, and bestow upon them such reality as they possess.*

This brings us back to Gibson's concept of "specification": the notion that information contained in the ambient optic array "specifies" objects and events in the environment. Here again we are confronted by an act of signification, an inherently *semantic* act which can be viewed as an immediate "presentification," an act that does not hinge upon a transfer of matter or energy through space. But how is that possible? That is a conundrum which neither our science nor our present-day philosophy are able to resolve. Something—some essential element—has evidently been left out of consideration: what might that be? It is none other, I say, than what Plato terms *eidos*, Aristotle calls *morphe*, and the Scholastics refer to as *forma*. Only "forms" can accomplish the prodigy in question: nothing of the kind, clearly, is to be found on the level of spatio-temporal realities. The fact is that to conceive of authentic perception one requires a notion of *morphe* or *eidos*: only a *form* is able to join a subject to an object so that "in a way" the two "become one" as Aristotle declares.

The crucial question, now, is how the requisite forms are given in the ambient optic array; and the answer is clear: they are given precisely in what Gibson terms *invariants*. Let us note that by the very conception of "invariance" these "entities" are allied to forms: for though an invariant may present itself in an indefinite number of spatio-temporal exemplifications, it actually transcends space and time and is consequently contained in no spatio-temporal thing or representation. Now, it is these invariants—these forms!—that permit perception. According to Gibson's theory, they are what is "registered" or extracted from the ambient optic array in the act of perceiving, and they are also what is objectively perceived. In a word, what bridges the chasm between "mind" and environment is none other than the invariants: they are in truth the *forms* that give access to the external world.

To proceed further we need to reflect upon the elusive notion of "consciousness." Let us pose the following question: Does the brain "produce" consciousness, or do we "become conscious" of what the brain is doing? If the brain be indeed a computer, does it generate consciousness, or does someone "read" that device? Either of these options has its protagonists; yet who can deny that both smack of absurdity: the idea of a "little man" reading the brain, surely, is no more credible than the claim that a computer—even one "made of meat"—can beget consciousness! One tends however to assume that there is no way out of this dilemma, no *tertium quid* to break the impasse. One fails to realize that the key to the problem—the decisive recognition that does break the impasse—is to be found in a perennial metaphysical teaching: the familiar tenet, namely, that "*the soul is the form of the body*," to put it in Aristotelian terms. That is what needs now to be explained.

We need to reflect upon the great and decisive fact that it is the *soul* (in the traditional sense of *psyche* or *anima*) that transforms the material substrate into a living and sentient body. And this means that the living body is *more* than what the molecular biologist or neurophysiologist conceives it to be. I have argued repeatedly that a *corporeal* object X, by virtue of its substantial form, is to be distinguished ontologically from the associated *physical* object SX, and that indeed the very atoms and molecules said to constitute X need to be distinguished, as parts of X, from atoms and molecules as conceived by the physicist; for indeed, as parts of X they partake somewhat of its substantial form. And that is the reason, I say, why elementary particles and their aggregates exhibit quantum indeterminacy: for it is by virtue of that so-called indeterminacy that these particles and their aggregates are capable of receiving an additional determination, which elevates them to the status of *bona fide* parts of a corporeal entity. My point, now, is that these considerations apply, *mutatis mutandis*, to the case of living and sentient organisms; it is only that in this instance the substantial form—traditionally referred to as a "soul"—belongs to a higher ontological order, and that consequently the ontological discrepancy between physical components and actual parts becomes correspondingly magnified. There is a world of difference, thus, between a neuron as conceived by the neurophysiologist, and an actual neuron in a living brain; and it is of course to be understood that in the absence of a soul there can be no sensations, no perception, no thought, and indeed no consciousness: without a soul there can be only protein molecules and potassium ions, artfully structured in ways the brain scientists are beginning to understand. But though consciousness does not derive from the molecules and ions that compose the physical brain, it is by no means extraneous to the living organism: there is no *homunculus* "reading the brain"! There need not be. The consciousness in question belongs neither to the material body nor to the soul as such, but to the living organism resulting from their union: it is a *psychosomatic* mode of consciousness, one can say (and to be sure, there *are* other kinds).

It is easy to understand that sensations pertain to the psychosomatic level of awareness; what is hard is to realize that *perception* does not. I contend that the perceptual act does not—and indeed cannot—take place on the psychosomatic plane. And why not? One way to argue the point is to note that it is implied by a remarkable feature of visual perception which Gibson was presumably the first to discern: the fact, namely, that perception does not take place in a temporal present, as had been surmised, but encompasses a duration, a span of time. The factor of movement, in particular, enters the picture, and not in a secondary capacity, but as an essential element, a *sine qua non* of perception. We need therefore to reflect upon the metaphysical significance of this fact.

To begin with, let me recall that traditional metaphysics rejects the idea of a temporal moment, the notion of an instantaneous temporal present. However, having banished the present from the flux of time, traditional doctrine reinstates that conception on a higher ontological plane. Yes, there *is* a "present"; but that present is not a temporal instant, not a present that "flows," but a *nunc stans* as the Scholastics say: a "now that stands." What needs to be grasped is that the act of perception—and in fact every cognitive act as such—takes place in a *nunc stans*, for the simple reason that temporal dispersion is inimical to the very essence of knowing. To know is perforce to know *one* thing, and this implies that one cannot know "in succession," piece by piece so to speak. One is right, therefore, in assenting to the common belief that perception takes place in a present, an indecomposable "now"; what is erroneous, on the other hand, is to conceive of that present in temporal terms as a "now" that moves. There *is* actually no temporal present: as the Scholastics recognized, the present is not a part of time.[15]

Now, the fact that the actual present is not in flux—is not indeed the temporal present of visual-image psychology—is precisely what renders possible the perception of stasis and change, of invariants and events. Gibson was right: we do perceive both persistence and alteration, and we do so without the intervention of memory. This fact, however, carries a deep implication which the scientist is prone to miss. The empiricist mind is able, certainly, to envisage a psycho-somatic domain; and Gibson, for one, has maintained that perception constitutes neither a physical nor a mental act, but pertains indeed to the psychophysical organism. One needs however to realize that the psychosomatic realm, by virtue of its somatic and hence material base, is subject to the temporal condition; in this realm "everything flows," as Heraclitus observed. But this implies that the *nunc stans*—and hence the act of perception—is not to be located in that domain. However "supra-temporal" the disembodied soul may be, the fact remains that, in union with the body, the soul becomes

15. Time is composed, not of moments, but of durations, even as a line is composed, not of points, but of intervals. As noted previously, it was Descartes who has misled us on that score.

subject to time. And this brings us, finally, to the essential point of these metaphysical considerations: the fact that perception takes place in a *nunc stans*, and thus "above time," entails that perceiving is not, strictly speaking, a psychosomatic act: the faculty by which we perceive proves not to be psychosomatic, but *spiritual*, and that spiritual faculty is what tradition terms *intellect*.

I will add in passing that what has no doubt impeded this recognition in the post-medieval West is the conflation of intellect with the faculty of reason: for inasmuch as the act of perception entails no reasoning—perception, in other words, is not inferential, but direct or immediate—it obviously cannot be assigned to the rational faculty. As regards perception in animals, it needs to be understood that, though admittedly bereft of rationality, animals too partake of intellect in some degree or mode. Even as all beings, no matter how apparently humble, participate in primary being, so too can it be said that all knowing participates in the primary intellect: one needs to realize that nothing in the universe is independent of the Center, nor has any reality apart from God.

Perception having now been identified as an *intellective* act, let me reiterate that sensation, on the other hand, is indeed psychosomatic: it is psychosomatic because it does clearly constitute a direct or unmediated response to sensory stimuli, which as Gibson points out, may be external, as in a sensation of light, or internal, as in the case of hunger or pain. Perception and sensation prove thus to be not only different, but correspond in fact to different metaphysical levels or planes; and this recognition places the Gibsonian discoveries in a new light. The claim, first of all, that perception is not sensation-based can now be supported on metaphysical grounds: a higher function can never be based upon a lower; that would be to invert the natural order. It is the lower functions, in fact, which are invariably dependent upon higher faculties, even as reasoning is dependent upon intellect, whereas the reverse is metaphysically impossible.

The greater the depth a science attains, the more it stands in need of metaphysical interpretation. It seems that with increasing depth incongruities bordering upon paradox make their appearance, to the consternation of the scientific community, which finds itself powerless to cope with such conundrums. We have already witnessed this phenomenon in the case of quantum theory, which "no one understands" without recourse to metaphysical discernment, beginning with the ontological distinction between the physical and the corporeal domains.[16] And now we encounter the same phenomenon in connection with the Gibsonian theory of visual perception: here too, I say, one comes upon incongruities that can only be resolved from a metaphysical point of vantage. Take Gibson's claim that "the world is not viewed in perspective," and that "one can perceive an object or a whole habitat at no fixed point of observation": what renders that contention quasi-paradoxical is the fact that it is inexplicable on a psychosomatic plane. Or take the tenet that change and motion are directly perceived, that is to say, perceived without the intervention of memory. What makes such recognitions incomprehensible not only to the layman, but to the scientist as such, is the fact that they are expressive of a metaphysical truth: the fact, namely, that *perception takes place in the intellect,* which is limited neither by space nor by time. It appears that in tenets such as these Gibson has isolated what may be termed "intellective" features of perception, traits that constitute effects of intellective agency which defy psychosomatic interpretation (just as state vector collapse, for example, defies *physical* explanation).[17] Let us not fail to note, moreover, that a psychosomatic act, by virtue of its somatic nature, is unable in principle to transcend the organism: if it be the case, therefore, that the terminus of visual perception is indeed ecological, as Gibson maintains, this in itself implies that the perceptual act is *not* psychosomatic. To put it plainly: if perception were psychosomatic, the percept could be no more external to the organism than a sensation of hunger or pain; and to be sure, no amount of "processing" can alter this fact. One sees that Gibson was justified in observing what amounts to total silence on the

16. See Chapters 2 and 3.
17. *The Quantum Enigma* (Tacoma, WA: Angelico Press/Sophia Perennis, 2012), chap. 6.

question how one *does* "register invariants": for indeed, there exist no psychosomatic means for the accomplishment of that feat, whereas the intellective are excluded *a priori* from scientific discourse. What to many of his colleagues appeared as a radical deficiency and mark of incompleteness in the Gibsonian theory proves thus to be its greatest merit: for that seeming deficiency is expressive of what constitutes in fact the very essence of the perceptual act.

The discernment, on empirical grounds, of what I have termed *intellective* traits of perception ranks, in my view, as one of the major achievements of twentieth-century science. These are the startling Gibsonian discoveries that have shaken the discipline of cognitive psychology, the seeming absurdities that have astounded all and scandalized many. It is a testimony to Gibson's scientific integrity that he refused to compromise on these issues, and a mark of genius that he was able to formulate a rigorous theory of visual perception, incorporating these seemingly incongruous tenets. It is of prime interest, moreover, that in so doing he has rectified the basic misconception endemic to our contemporary Weltanschauung: the misbegotten notion of Galileo and Descartes that the terminus of perception is to be located in the mind or brain of the percipient. What Gibson has left us as his legacy is sound empirically-based science that can indeed be integrated into higher orders of knowledge, as we have come to see.

5

NEURONS AND MIND

"The Astonishing Hypothesis," writes Sir Francis Crick in a book bearing that title, "is that 'You,' your joys and sorrows, your memories and your ambitions, your sense of personal identity and free will, are in fact no more than the behavior of a vast assembly of nerve cells and their associated molecules."[1] To be sure, notions of this kind have been expressed time and again since the days of Democritus; what is new, however, in the case of contemporary neuroscience, are the grounds in support of the reductionist contention. Over the past century a vast and altogether unprecedented body of knowledge concerning the structure and functioning of the human brain has come to light, which does seemingly justify the hypothesis in question. For example, technologies have evolved which permit scientists to discern the actual firing of neurons in the living brain, thus allowing them to track the correlation of neural firing patterns with the conscious mental life of the subject. Or take pharmacology: that science has currently advanced to the point where one is able to produce "mind-altering" drugs by tailoring molecules to interact with neurochemical substances in specific ways. We need not belabor the point: what Crick terms "the scientific search for the soul" has now begun in earnest, and on a stupendous scale.

I propose in the present chapter to reflect upon this scientific quest in light of sacred tradition. What stands at issue is not simply the truth or falsity of Crick's Astonishing Hypothesis, but above all, a comprehension of how "mind" does relate to neural function.

---

1. *The Astonishing Hypothesis* (New York: Simon Schuster, 1995), p. 3.

What is called for, once again, is a separation of scientific facts from scientistic misconceptions, and a certain integration of the former into metaphysical orders of knowledge; only this time the stakes are higher than ever before: the contest now is for the *soul*.

It was not till the twentieth century that the nerve cell or neuron was identified as the fundamental component of the nervous system by the Spanish anatomist Ramón y Cajal. One knows today that a single layer of the cerebral cortex contains approximately 100,000 of these cells per square millimeter, and that about 100 billion neurons are needed to make a human brain. Admittedly, the fact that the brain is "made of neurons" does not mean that its operation can be understood even in principle by way of neurophysiology; it does however imply that one can *not* understand the brain without first understanding the anatomy and physiology of neurons.

A neuron may be broken into three components: soma, dendrites and axon. The soma is the central body of the cell containing the nucleus with its approximately 100,000 genes. The dendrites constitute a treelike network of fibers growing out of the soma, whose function it is to receive incoming signals. The axon, finally, which constitutes the "output cable" of the neuron, consists of a central nerve fiber, which may in some cases be several feet in length, and which typically branches near its outer extremity. All these axonal or efferent nerve fibers (and occasionally a few of the dendritic fibers as well) terminate in a bulb-like structure known as a synaptic knob, which controls the transmission of electrical signals to neighboring cells. It is important to understand that this transmission is accomplished and regulated by chemical means through the secretion of substances, known as neurotransmitters, into the so-called synaptic cleft. Synaptic knobs and their neurotransmitters may be excitatory or inhibitory, and a recipient neuron reacts to a kind of algebraic sum of excitatory and inhibitory electrochemical signals generated by nearby nerve cells. The process is exceedingly complicated and constitutes one of the many prodigies of molecular design which have come to light in recent decades. Suffice it to say that an understanding of brain function hinges upon a detailed knowledge of this molecular mechanism, a domain of neuroscience which today is finding applications in pharmacology and medicine.

Having noted that a nerve cell responds to "a kind of algebraic sum" of incoming signals, let me now say a few words concerning the generation and transmission of this electrical response within a given neuron. We may think of a nerve fiber as a cylindrical tube containing an ionized solution of sodium and potassium chloride, separated from a similarly constituted ambient fluid by a membrane. An excitatory stimulus produces a positive potential spike near the base of the axon; and when this positive potential reaches a certain threshold, it activates a molecular mechanism in the membrane consisting of so-called gates: there are potassium gates, which cause potassium ions to move from inside the axon into the ambient fluid, an action which flattens the spike (remember, potassium ions are positively charged), and there are sodium gates which pump sodium ions in the reverse direction and have an opposite effect (sodium ions likewise carry a positive charge). These respective actions are coordinated to move the potential spike outwards along the axon, and can do so at speeds up to 300 feet per second. It is noteworthy that there is no motion of electric charges in the direction of transmission, and no driving potential difference between the terminals, as is the case in man-made electrical devices. It appears that this marvel of nanotechnology is to be found in neurons throughout much of the animal kingdom, down to the invertebrates. I should point out that since the potential spikes produced by a neuron all have the same "algebraic sign" and the same amplitude, the only parameter that is variable and thus carries information is their frequency or time distribution. In the absence of appreciable stimuli a neuron tends to fire sporadically at a low background frequency, roughly between 1 and 5 Hertz; when stimulated to the threshold level, on the other hand, its frequency increases sharply (typical firing frequencies in excited neurons range between 50 and 100 Hertz, and can at times approach 500). Finally, it should be mentioned that there are many different kinds of neurons in the brain, each exhibiting its own special characteristics in conformity to its function.

Following this introductory statement regarding the nature of neurons, I propose to consider next the division of the human brain into regions associated with various recognizable functions. Never mind, for the moment, how these functions are to be explained in terms of neuronic interactions; what will concern us in the following section is simply the functional geography of the human brain.

The major anatomical divisions of the brain can be discerned through dissection and have been known for a very long time. The largest and uppermost portion, as everyone knows, is the cerebrum, which is split down the middle into left and right cerebral hemispheres, and crosswise into four lobes: frontal, parietal, temporal and occipital. It further divides into an outer and inner layer, known as the cerebral and cerebellar cortices, which correspond to the so-called grey and white matter, respectively.[2] Tucked beneath the occipital lobes near the back of the head lies the cerebellum or "little brain," which Darwinists tend to look upon as representing the brain of our distant mammalian ancestors. Besides the cerebrum and cerebellum, there is a major group of brain components, known as the limbic system, partly hidden within the central cavity beneath the cerebrum, which comprises the hippocampus, thalamus, hypothalamus and amygdala. Below these formations lies the brain stem, which somewhat resembles the brain of reptiles, and is generally thought to have evolved "more than 500 million years ago." There are yet other components—even the retina is nowadays regarded as part of the brain—but let this suffice.

It appears that in addition to its anatomical divisions, the brain admits functional divisions as well. To be more precise: there exist functional modules which can be localized anatomically, at least in some rough way. Basically this amounts to saying that different parts of the brain do different things. Neuroscientists, understandably,

2. The "white" of white matter is due to a substance called myelin found in the coating of longer axons which increases the speed of neural transmission. The cerebellar cortex appears white because it is mainly composed of axons.

have been hard at work to determine "what is done where," an enterprise sometimes referred to as mapping the brain.

At the risk of a slight digression, let me begin this brief survey with a reference to Franz Josef Gall, the founder of phrenology, who tried two hundred years ago, with extraordinary ingenuity, to map the brain by mapping the skull; the end result was a kind of cranial atlas labeled by functional terms. As might be expected, there was a region corresponding to "Amativeness," and another associated with Combativeness, a zone which Gall had identified on the basis of its smallness in "most Hindoos and Ceylonese"! It appears that the good doctor got lucky at times, for example, when he stipulated a region of Mirthfulness within the left temple. Two centuries later, surgeons at the UCLA Medical Center, probing the cerebral cortex of a patient by means of localized electrical stimuli, were surprised when the young woman (who was fully conscious) suddenly burst into fits of laughter: it appears they had indeed hit upon a "region of Mirthfulness" in the left frontal lobe! Questioned about the cause of her mirth, the woman replied: "You guys are just so funny—standing around." This is exactly the kind of response brain-mapping scientists wanted to hear.

Before the advent of modern medical technology, the prime scientific means of mapping the brain was to correlate loss of function with brain damage, the location of which could be determined postmortem through autopsies. There is the famous case of Phineas Gage, a young laborer in Vermont, who in the year 1848 had an iron bar a yard long driven through his brain by an explosion. Amazingly, Gage survived, and in fact could live a biologically normal life; what was missing, however, was the ability to control his impulses and direct his actions towards normal goals. It appears that the centers associated with these "higher" functions were localized in portions of his frontal lobes that had been permanently destroyed. Another early example of functional localization is the discovery by Pierre Broca and Carl Wernicke of so-called language areas, which to this day bear their names. Both are normally located in the left cerebral hemisphere; Broca's area has to do with speech formation and is situated in the frontal lobe, and Wernicke's area has to do with the comprehension of speech and is situated in the temporal.

For various reasons, interest in the mapping problem waned during the early decades of the twentieth century, and it was due in part to the remarkable inquiries of Wilder Penfield, the Canadian neurosurgeon, that the field became active again in the 40's and 50's. Penfield studied the brain of conscious patients, laid bare through the surgical removal of the upper skull, and was able, by this means, to obtain a wealth of accurate information. Meanwhile scanning and tomography have come into play, and one is now able to view not only brain structure, but brain activity as well. To mention at least one such technology: what is known as functional Magnetic Resonance Imaging (fMRI) can deliver up to four images per second, which is fast enough to "film" the large-scale effects of neurological action associated with conscious activities. In this way one can "look in" upon the living brain, and conduct psychophysical observations with comparative ease.

As one might expect, numerous regions within the brain have so far been "identified," and many more are under scrutiny. Within the cerebral cortex, for example, neuroscientists have located both sensory and motor regions, which have moreover been broken down into primary, secondary and tertiary domains, and into even finer subdivisions. One speaks of "recognition pathways," and of recognition units or RU's which can be immensely specific. A single brain lesion, for instance, can destroy one's ability to recognize a human face without impeding the capacity to recognize other things, including animals, a condition known as prosopagnosia. A remarkable case in point is that of a farmer who became incapable of recognizing his friends as a result of an injury, but could recognize each of his thirty-six sheep and call them my name. Other kinds of functional units associated with still higher levels of mental activity seem to exhibit corresponding degrees of specificity. Scientists at the University of California, for example, have even identified an FU which supposedly is specific to religious or mystical experience, and is said to produce "intense feelings of spiritual transcendence, combined with a sense of some mystical presence" when stimulated.[3] I will mention in

---

3. Rita Carter, *Mapping the Mind* (Berkeley, CA: University of California Press, 1999), p. 13.

this connection that experiments involving practitioners of yoga have shown that yogic forms of meditation can "turn off" areas of the parietal and premotor cortices which are normally active.

A great deal is known about the function of non-cerebral regions within the brain as well, beginning with the limbic system. Speaking in highly general terms, one knows that the hippocampus is involved in the formation of long-term memories, and that the hypothalamus controls various emotions and drives, such as hunger. The amygdala has sometimes been described as the body's alarm system, and is also a major player in the emotional life; it is implicated, for example, in the formation of phobias. The sensation of stark terror which many people experience at the sight of a coiled or slithering serpent, for example, seems to originate in that particular portion of the brain.

There is much evidence to support the currently popular notion to the effect that the "left brain" is rational and analytical whereas the "right brain" is intuitive and operates in a more holistic manner. The two cerebral hemispheres are normally connected by a neural bridge known as the corpus callosum, across which messages are transmitted in both directions along some 80 million axons. In the 1940's it became medically fashionable in cases of severe epilepsy to sever this neural connection surgically, a procedure known as lobotomy, and it is estimated that more than twenty thousand lobotomies were performed in the US alone. The procedure was pioneered by a Portuguese neurologist named Egas Moniz, who discovered that he was able to pacify aggressive chimpanzees by cutting nerve fibers in their frontal lobes. I might mention that he was eventually shot and killed by one of his lobotomized patients. It appears that the unfortunate victims of this monstrous procedure are indeed split personalities, irreparably mutilated. Roger Sperry, a psychobiologist who worked extensively with lobotomized patients (for which he received a Nobel Prize) tells us that "Everything we have seen indicates that the surgery has left these people with two separate minds." I will conclude this section with a case history which seems to support this conclusion.

The patient, identified as P.S., was subjected to experiments by two neuroscientists, Joseph LeDoux and Michael Gazzaniga.

Whereas most people have no language abilities in the right cerebral hemisphere, it happens that P.S. had acquired a rudimentary ability to understand simple phrases and to communicate by means of words in that region of his brain. LeDoux and Gazzaniga wanted to utilize this rare capacity to interrogate both hemispheres independently. The questions could not be posed verbally, because, unlike visual images, sounds cannot be communicated to one hemisphere without the other "listening in": the way the auditory nerves are connected to the brain makes this impossible even in lobotomized subjects. I will let Rita Carter, a science journalist, pick up the story from here:

> LeDoux and Gazzaniga got around this problem by presenting P.S. with spoken phrases and questions minus key words that would make them answerable. This essential information was then sent to the right hemisphere only by presenting the key words visually. Thus they might say "Please would you spell out..." and then flash the word "hobby" in his left visual field (which goes to the right hemisphere). This convoluted exercise insured that the right hemisphere was the only half with all the information required to formulate a reply. P.S.'s right hemisphere could not generate speech, but it was able to write. It therefore spelt out its answers, using P.S.'s left hand (the one under right-brain control) to organize Scrabble letters into words.[4]

The results were surprising. In response to the question "What do you want to do when you graduate?" the left hemisphere stated its wish to become a draftsman, whereas the right hemisphere gave out (via Scrabble) that it wanted to be a race-car driver! Roger Sperry was right: it does appear that P.S. had "two separate minds."

---

4. *Mapping the Mind*, op. cit., pp. 50–51.

We have been concerned, up to this point, with functional units and their localization within the brain, without reference to their internal structure and rationale of operation. In contrast to this "black box" approach, we propose now to consider what the brain actually does: what transpires *inside* the black box. Neurologically speaking, the generic answer is given in advance: neurons interact. They receive stimuli and in turn stimulate by firing; that is all that happens, all that *can* happen in a system composed of neurons. What matters are massive firing patterns involving a vast number of nerve cells; a handful of neurons amount to nothing at all so far as conscious experience or motor control are concerned. A neuron is structured for interaction, designed for membership in a community of similar units.

The analogy with transistors is obvious: these man-made devices too have input and output junctions, and are designed to interact with each other within a network. There is consequently an analogy as well between the brain or its FU's and man-made computers, and it is hardly surprising that this analogy has served as a major source of inspiration in the neuroscience and AI communities.[5] It has tempted many—beginning with Alan Turing himself[6]—to surmise that computers can not only simulate mental processes, but can in principle generate such, a tenet which carries the label "strong AI." It should be noted that strong AI is actually stronger than Crick's Astonishing Hypotheses, which merely reduces mind to "the behavior of a vast assembly of nerve cells and their associated molecules," without prejudging whether the brain functions simply as a computer or so-called Turing machine. In the early days of AI euphoria,

5. AI stands for "artificial intelligence," the discipline concerned with devices that simulate or manifest intelligent action.

6. Alan Turing was possibly the foremost computer theorist, a man of uncanny logical and mathematical intelligence, who worked for the Allies in World War II as the leading code breaker. He conceived of what is termed the Turing machine, a formal device which constitutes the prototype of every actual or possible computer. He is also known for his conviction that the human mind itself is a Turing machine. Tragically, his life ended in suicide.

it was not uncommon to assume that the brain functions in fact like a von Neumann computer: a special kind of Turing device based upon serial as opposed to parallel processing. One knows today that such an arrangement would be biologically unfeasible for a number of reasons, beginning with the fact that neural action is far too slow to permit efficient serial operation: neurons fire at a maximum rate on the order of 500 spikes per second, which is about a million times slower than the "firing rate" in a respectable computer.

Among the salient differences between the brain and a computer one might mention, first of all, the fact that neurons generally have a vast number of input and output connections, as compared to a mere handful in a transistor (some neurons have up to 80,000 synaptic connections). Moreover, as has often been pointed out, there appears to be considerable randomness and redundancy in neural connections; one needs but to examine a sample of brain tissue under a microscope to realize that it lacks the structural regularity of a computer component. There is some randomness too in neural response to stimuli, apart from the fact that the brain obviously does not operate on a binary system. Moreover, synapses are known to have different and indeed variable characteristics, a feature which turns out to play a vital role. In addition, the brain can grow new synapses by way of protrusions known as dendritic spines, a phenomenon referred to as plasticity. If it be indeed a computer, it is one that knows how to "rewire" itself.

Despite these fundamental differences, however, the fact remains that there is an analogy between the brain and a computer which can be applied to great advantage by the neuroscientist. A certain grasp of computer science appears in fact to be required for even a rudimentary understanding of brain function; it is no coincidence that with the advent of computer technology, neuroscience has experienced a second birth. Admittedly, the hopes and expectations that had prevailed in the days of von Neumann have not been realized, and the neuroscience community has meanwhile become more cautious in its "algorithmic" claims; yet, even so, highly significant progress has been made in the application of computer-related concepts and techniques to the scientific understanding of the brain. To cite at least one example: computer programs known as neural networks have been applied with considerable success as a means of simulating various kinds of neurological processes, and have led to some remarkable insights concerning the manner in which the brain executes certain tasks.

One of the first enigmas to give way pertains to the formation of memory. The key notion was supplied early on by a Canadian psychologist named Donald Hebb; the idea, as he put it, is this: "When an axon of cell A is near enough to excite a cell B and repeatedly and persistently takes part in firing it, some growth process or metabolic change takes place in one or both cells such that A's efficiency, as one of the cells firing B, is increased." This principle, which has come to be known as Hebb's rule, appears to provide the basis for certain forms of memory and learning. Though the factors which regulate the ability of one neuron to stimulate another are not yet fully understood, it is known that these do include growth processes (as in the case of dendritic spines) as well as chemical changes affecting the vicinity of a synaptic knob. Let us suppose that a large group of neurons have bonded in a Hebbian manner; it may be possible, then, to trigger an extensive firing pattern, representing the content of a memory, by means of a much smaller firing pattern, representing what is termed a clue. This conjecture has in fact been verified with the aid of a neural network proposed in 1982 by a brain scientist named John Hopfield.[7] The network consists of units representing neurons and connections representing synapses, endowed with "weights" representing the relative strengths of these synaptic connections. Each unit has one output and a number of inputs, and the network is wired to feed back upon itself, so that the output of one cycle becomes the input of the next. It turns out that given an arbitrary initial input, the system will eventually converge to a stable output. Moreover, if even a small part of the resultant pattern (corresponding to a clue) is supplied initially, the system will converge to the given pattern in just a few cycles. "As a result," observes

---

7. Like Sir Francis Crick, Hopfield is a physicist turned brain scientist by way of molecular biology. It seems that in the course of the last century the "center of scientific interest" has actually shifted from physics to molecular biology and neuroscience.

Crick, "it will, in effect, have produced 'memory' from something that merely nudged its memory." As Crick goes on to explain:

> Notice that the "memory" need not be stored in an active state but can be entirely passive, since it is embedded in the pattern of weights. . . . The network can be completely inactive (with all outputs put at O), yet when a signal is fed in, the network will spring into action and in quite a short time settle into a state of steady activity corresponding to the pattern that had to be remembered. It is surmised, on good grounds, that the recall of human long-term memory has this general character.[8]

What is more: a Hopfield network can "remember," not just one, but many distinct patterns, each of which can be "recalled" by means of a corresponding clue, as is the case in human long-term memory. In sum, memory is distributed over many connections, memories are superimposed (because a single connection can enter into many memories), and most importantly, from a biological point of view, memory is robust, since it is effected by a mass action which is not sensitive to the behavior of a relatively small number of neurons (we can lose hundreds of neurons per day without noticeable effect upon our memory).

No one claims—or ought to claim—that we carry a Hopfield network in our brain. What Hopfield's investigations show is that a long-term and "content addressable" memory can be explained on a Hebbian basis, whether it be by means of a neurological network of the Hopfield variety or in terms of some other network exhibiting the same general characteristics. Whatever the case may be, it appears that neuroscience has begun to unravel the enigma of long-term memory.

No functional system within the brain has received greater scrutiny than the visual. Much of this research has been carried out on animals, and especially on the macaque monkey, whose visual system

8. *The Astonishing Hypothesis*, op. cit., p. 184.

appears to be closely similar to the human. The need for animal subjects arises primarily from the fact that long neural connections are studied by injecting chemicals into the brain and tracking their transport; it is necessary, therefore, within hours or days of the injection, to sacrifice the subject, in order that brain tissue can be examined microscopically before the chemicals have too much dispersed. Sadly, thousands of animals have thus been put to death in the interest of neuroscience. Getting back to the visual system: its complexity—say, in the macaque—can hardly be imagined; the most complicated wiring diagram ever devised by a team of engineers is miniscule by comparison. I wish now to speak of the human visual system in suitably rough terms, and will assume (as neuroscientists commonly do) that what has been learned from the macaque carries over more or less to the human brain.

The system divides, first of all, into a number of subsystems, all of which receive inputs from retinal neurons. Each subsystem, obviously, has its own special function; what is termed the secondary system, for example, appears to be principally concerned with the control of eye movements. Leaving aside all such "auxiliary" subsystems, we shall restrict our attention to the so-called *primary* visual system, the subsystem most directly responsible for visual perception. The story begins in the retina, which houses four kinds of photoreceptors: rods (more than 100 million in each eye), whose primary function it is to respond to dim light, and three kinds of cones, each of which responds to a different range of wavelengths. The reader will note that it is by virtue of these specificities that we possess both night and color vision. The primary visual system projects the resultant retinal output to an organ in the thalamus known as the Lateral Geniculate Nucleus (LGN) via neurons known as ganglion cells. We may think of the LGN as a gateway or relay station for nerve impulses on their way to the higher visual areas in the cerebral cortex, beginning with an area designated as V1. Some twenty cortical visual areas have thus far been identified. All these regions (including the LGN) are stratified, and are usually divided into six layers. The neurological connections between different layers in the same region, and again, between distinct regions, exhibit certain rules, of which the most intriguing seem to be those that concern layer 4. To oversimplify a bit: One can distinguish "forward" from "back" projections between distinct visual areas on the basis of whether these connections do or do not feed into layer 4. The retina, in particular, projects to the LGN, and the LGN projects to V1, by forward projections. Continuing by way of forward projections, one obtains a hierarchic ordering of the visual centers, in which V1 is followed by V2, and so forth, a sequence which terminates in the hippocampus. There are complications, on account of which the experts speak of a "semihierarchical" order; but the simplified picture we have drawn will suffice for the purpose of this overview.

Since every neuron in the visual system is connected to the retina by neural pathways, one can speak of its "receptive field" as that portion of the retinal surface within which it can respond to stimuli. Consider now a particular layer within a given visual area; so long as the receptive fields of the neurons involved are sufficiently small, the neural connections define a kind of map or point-to-point correspondence between a portion of the retinal surface and a corresponding region of that particular layer, a fact which permits one to speak of retinoptic maps. Lest it be thought, however, that a retinoptic map is something to be "looked at"—as if by a "little man" within the brain—let me point out immediately that the notion of retinoptic map ceases to apply as we ascend into the higher visual areas, due to the fact that the receptive fields tend to become large and can indeed cover the entire visual field of the single eye.

We come now to a major key, a recognition that proves to be essential, which came to light through a series of experiments conducted by David Hubel and Torsten Wiesel, for which they received a Nobel Prize in 1981. They began in the late 50's to record electrical impulses from single cells in the V1 visual area of cats by means of microelectrodes; to everyone's surprise, they found that the neurons responded not simply to "light or dark" within their receptive field, but in feature-specific ways. One class of cells, for example, responds most strongly to lines or edges, and to a preferred orientation of these visual elements. Some cells seem interested in short lines, others in longer lines; some respond to the position of a pattern, others to its motion. There are cells which fire best to a particular direction of motion, and on higher levels of the visual system there are cells which seek out the motion of an object relative to its background. As a rule, the higher one goes, the more selective and sophisticated the neurons tend to be. For example, V2 already contains neurons which fire to so-called subjective contours: to certain lines, namely, which are perceived but are not given in the retinal image. One of the most fascinating visual areas thus far uncovered is V4, which is concerned with the perception of color. It has been known for some time that the color we perceive is not simply a function of the wavelength, but depends in complicated ways upon other factors as well; for example, the color of a particular patch in the visual field is affected by the colors of neighboring patches, a fact known as the Land effect. Now, in an experiment utilizing this effect, it has been shown that cells in the macaque V4 region which normally fire to the color red continue to do so even when the actual wavelength has been altered: it was found that the cell fired whenever the experimenter himself perceived the given patch as red.

It appears that Hubel and Wiesel have uncovered a major characteristic of the visual system: the visual hierarchy, one finds, is geared to the recognition of ever more complex features latent in the initial input. As yet the "logic" of the system is poorly understood. For example, one still knows little about the function of back connections, concerning which there has been a rather wide range of speculation. The overall structure of the system, however, is no longer in doubt; as Crick points out:

> The general pattern, then, is that each area receives several inputs from lower areas. . . . It then operates on this combination of inputs to produce even more complex features, which it then passes on to even higher levels in the hierarchy.[9]

The process, clearly, is analytical; at each level the input is broken down into components of some kind. One might compare a visual area to a filter which passes or blocks out inputs in accordance with

---

9. *The Astonishing Hypothesis,* op. cit., p. 158.

criteria of its own. The information which is given synthetically in the retinal input comes thus to be spread out over various fields, each answering to its own set of parameters. One might think that there must be a "final field" which corresponds to what is actually perceived, and is presumably to be found at the summit of the visual hierarchy; but it happens that one can lose all the visual areas above a certain level and still see quite well. "In short," Crick observes, "*we can see how the brain takes the picture apart, but we do not yet see how it puts it together.*"[10] The point, however, as we shall learn in the end, is that the brain *does not in fact "put it together" at all.*[11]

Clearly, the question Crick has raised is not special to visual perception. "As neuroscientists keep subdividing the brain," writes the science journalist John Horgan, "one question looms ever larger. How does the brain coordinate and integrate the workings of its highly specialized parts to create the apparent unity of perception and thought that constitutes the mind?"[12] This is the conundrum that

10. Op. cit., p. 159. Crick's italics.

11. It is to be noted that this supports a central contention of James Gibson (see Chapter 4) to the effect that visual perception cannot be explained on a neurological basis. I should perhaps point out that when Gibson was formulating his ideas in the 50's and 60's, neuroscience was just beginning to uncover the basic neurological facts relating to the structure and functioning of the visual system. It seems that Gibson, psychologist that he was by training and professional interest, was not at the time in a position to assimilate these findings to the point of recognizing their relevance to his own research. Today, with the benefit of hindsight, we can readily understand that neuroscience has corroborated many Gibsonian tenets, beginning with his revolutionary claim that we do *not* perceive a so-called visual image, be it retinal or retinoptic. For not only are there many different kinds of retinoptic maps, none of which correspond to what we actually see, but what is more, retinoptic maps become increasingly distorted as we move into the higher visual areas, which eventually cease to be retinoptic, cease to be "maps" of the visual field. Another Gibsonian contention which has now been confirmed on neurological grounds is that what is perceived need not be given directly in the form of retinal stimuli: the fact that neurons in some of the higher visual areas can fire to so-called subjective contours corroborates that conclusion.

12. *The Undiscovered Mind* (NY: Simon & Schuster, 1999), p. 22.

has come to be known as the binding problem; ignored or waved aside by many, it has exercised some of the best scientific minds of our time. One of these is Roger Penrose, the Oxford professor of mathematics who, in 1970, proved the celebrated singularity theorem for black holes in collaboration with his student, Stephen Hawking. He has since turned his attention from the macrocosm to the microcosm, and has become deeply involved in the study of the human brain. As might be expected, Penrose began his inquiries by investigating the implications of the computer paradigm. He was in effect testing the strong AI hypothesis to ascertain whether that premise can in principle explain the phenomenon of human thought. From the start he positioned himself on the highest level of generality, namely that of Turing machines, thereby obviating the need to distinguish between serial and parallel processing devices. Through an ingenious application of what is commonly referred to as Gödel's theorem, Penrose was able to show that the mathematical mind has the capacity to solve problems which cannot, *in principle*, be solved by computational means. What computers do, obviously, is to compute; it turns out, however, that the most characteristic operations of the human mind are not in fact computational or *algorithmic*, to use the technical term. Take the ability to distinguish between truth and error: "Indeed," says Penrose, "algorithms, in themselves, never ascertain truth!" Even in the case of problems supposedly "solved" by computers, the mathematician proves to be indispensable; it is he, after all, who programs the computer, and it is he again who interprets its output.

Nonetheless, inherently algorithmic operations, carried out in the brain, do evidently play an essential role in human thought: the very structure of the nervous system apprises us of this fact. Early on, however, Penrose reached the conclusion that the computer paradigm applies primarily in the unconscious realm: "The hallmark of consciousness," he maintains, "is a non-algorithmic forming of judgments." Once again, it is in the sphere of mathematics that Penrose arrived at this inference:

> One needs external insights in order to decide the validity or otherwise of an algorithm.... I am putting forward the argu-

ment that it is this ability to divine (or 'intuit') truth from falsity (and beauty from ugliness!) in appropriate circumstances that is the hallmark of consciousness.[13]

Repeatedly Penrose emphasizes this cardinal and inherently Platonist point: "We must 'see' the truth of a mathematical argument to be convinced of its validity. This 'seeing' is the very essence of consciousness."[14]

Penrose was aware of the fact that the conception of mathematical judgment at which he had arrived stands in opposition to the customary textbook version of what it means to be rational: "Often it is argued," he writes, "that it is the conscious mind that behaves in a 'rational' way that one can understand, whereas it is the unconscious that is mysterious." I would only add that in recognizing that the mathematician—of all people!—does not "behave in a 'rational' way that one can understand," Penrose has rediscovered what in traditional parlance is termed "*intellect.*" His inquiries have disproved the notion that we can somehow calculate our way to knowledge and understanding, as modern man has been taught to believe. By recognizing that rationality as such—the very thing which supposedly could dispel all mystery—is itself profoundly mysterious, Penrose has reopened the door to authentic metaphysical discovery. I should emphasize that what is "mysterious" in the sphere of mathematics is not indeed the formal content of a theorem or its formal proof, but the *seeing* of that content and the *seeing* of that proof.

In seeming opposition, moreover, to what Penrose himself has told us concerning the hallmark of consciousness, it appears that the best mathematics is sometimes accomplished beyond the pale of our ordinary or "individual" consciousness. In this connection Penrose relates an experience in the life of the mathematician Henri Poincaré: having interrupted his mathematical investigations regarding so-called Fuchsian functions to go on a geological excursion, Poincaré was boarding a bus when suddenly, without any

---

13. *The Emperor's New Mind* (Oxford University Press, 1990), p. 412.
14. Op. cit., p. 418.

connection to what was going through his mind at the time, he was struck by the idea "that the transformations I had used to define the Fuchsian functions were identical with those of non-Euclidean geometry." This recognition, which turned out to be correct, proved to be crucial. Penrose himself, moreover, has a similar story to tell: the essential idea which underlies the aforementioned singularity theorem came to him as he interrupted a conversation about other things with a visiting colleague to cross a busy London street. It is highly significant that Penrose refers to an intense "elation" to which this momentary event gave rise, and which in fact enabled him, after his colleague had departed, to search his memory and recover the idea in question; we shall have occasion in the sequel to comment upon this remarkable incident.

It is hardly surprising that Penrose became deeply interested in the binding problem. John Horgan, the science journalist, had occasion to talk with him on that subject in 1994, when representatives of various disciplines gathered in Tucson to attend a conference having to do with the nature of consciousness: "Toward a Scientific Basis of Consciousness" was the conference title. I might mention, in passing, that there was no dearth of colorful personalities in attendance. Danah Zohar, for example, "who earned a degree in physics from MIT and then studied philosophy and religion under the psychoanalyst Erik Erikson at Harvard," was on hand to expound views previously expressed in her 1990 book, *The Quantum Self.* David Chalmers, an Australian philosopher, was there to expound his own version of strong AI: "According to this theory, any object that processes information must have some conscious experience." Even a thermostat, it seems! Present as well was Christof Koch, a major figure in the world of neuroscience, who spoke about synchrony in neural firing and the significance for consciousness research of the so-called 40 Hertz frequency; and Walter Freeman from Berkeley, who promoted the notion that consciousness has something to do with chaos theory. But let us get back to the distinguished Oxford mathematician and the binding problem: "Penrose concluded," writes Horgan, "that no mechanical, rule-based system—that is, neither classical physics, computer science, nor neuroscience as currently construed—can account for the mind's creative capacity."[15] As Penrose himself stated at the Conference: "*What computers can't do is understand.*" It appears, however, that once again Penrose was on the spoor of an idea: "He now suggested," Horgan continues, "that quantum nonlocality might be the solution of the binding problem." What is called for, Penrose believes, is what he terms a "correct quantum gravity" theory, a kind of physics which does not yet exist. Now, for my part, I very much doubt that a new physics will one day unravel the mystery of consciousness. I do believe, however, that Penrose was right in suggesting that the solution of the binding problem hinges upon "quantum nonlocality"; it is only that the metaphysical and indeed ontological implications of that nonlocality need to be recognized. The crucial point, as I have argued elsewhere,[16] is that nonlocality refers in truth to the intermediary or "subtle" domain, the *bhuvar* of Vedantic cosmology, which is not subject to the spatial condition. It is thus in the "subtle body"—the Vedantic *sūkshma-sarīra*, and not in the brain!—that the elusive "binding" takes place. This is what I shall now attempt to explain.

I propose thus to approach the "mind-body" question in terms of Vedantic anthropology. It will not be possible, obviously, to expound the latter doctrine even in summary fashion within the compass of this chapter;[17] what I shall do is to introduce the key concepts in terms of which an answer to our query can be framed. One needs in the first place to understand that, according to Vedanta, man possesses, not just one, but *three* "bodies," corresponding to the three principal degrees of manifestation[18]: the

15. Op. cit., p. 240.
16. "Bell's Theorem and the Perennial Ontology," in *The Wisdom of Ancient Cosmology* (Oakton, VA: The Foundation for Traditional Studies, 2004).
17. The definitive treatise on that subject, in a European language, is doubtless René Guénon's *Man and His Becoming According to the Vedanta*, to which I refer the interested reader.
18. It should be noted that this corresponds to the triadic division *corpus-anima-spiritus* of the Western tradition.

sthūla-śarīra or "gross" body, the *sūkshma-śarīra* or "subtle" body, and the *kārana-śarīra* or "causal" body. There is also however another triadic division that needs to be recalled: a division, namely, of the *sūkshma-śarīra* itself, said to be comprised of *kośas*, the so-called sheaths or "coverings" of Purusha, the indwelling Self. According to this doctrine the *sūkshma-śarīra* breaks into a *prāṇamaya-kośa*, a *manomaya-kośa*, and a *vijñānamaya-kośa*, that is to say, into a sheath "made of *prāṇa*," a sheath "made of *manas*," and a sheath "made of *vijñāna*." It is of course impossible to find exact equivalents of these Vedantic terms in Western languages; in a general way, however, *prāṇa* corresponds to the vital force or *élan vital*, *manas* to mind, and *vijñāna* to a higher cognitive faculty answering to the traditional conception of *intellect* (the Sanskrit term *jñāna* being cognate to the Greek *gnosis*). It is of interest to note that a corresponding division of the soul or *anima* is to be found in the Western tradition, and that St. Thomas Aquinas, in particular, distinguishes between a *vegetative*, a *sensitive*, and an *intellective* soul, based upon a distinction of powers, which in turn derives from a distinction of the corresponding objects; as St. Thomas explains:

> But the object of the soul's operation may be considered in a triple order. For in the soul there is a power the object of which is only the body that is united to that soul; the powers of this genus are called vegetative.... There is another genus of powers in the soul, which regards a more universal object, namely, every sensible body, not only the body to which the soul is united. And there is yet another genus of powers in the soul, which regards a still more universal object, namely, not only a sensible body, but universal being itself.[19]

Though the Vedantic and Thomistic doctrines evidently represent different points of view—different *darśanas* as Hindus might say—one sees that the Thomistic criteria apply to the *kośas* of the subtle body as well. Let us consider first the *prāṇamaya-kośa*, which does indeed correspond to the vegetative soul. Its function, one might

19. *Summa Theologiae*, Quest. 78, Art. 1.

say, is to bond with the gross or corporeal body, and thus to act as the intermediary between the latter and *manas* or mind. It is needful, however, to note that the former exists as an actual body or *śarīra* precisely by virtue of this fusion with the pranic sheath. It is consequently imperative to distinguish categorically between the body conceived as a corporeal entity X and the living body LX which constitutes the outermost sheath of the integral human being. It is to be noted that a transition from LX to X takes place at the moment of death, when the *prāṇamaya-kośa* is withdrawn, leaving the "merely corporeal" body behind. Thomistically speaking, the resultant body has no longer a substantial form, and has thus become reduced to a mere composite or mixture of corporeal substances, which as such is subject to decay. It is noteworthy, moreover, that the Vedanta refers to the gross body as *annamaya-kośa*, the sheath "made of food": what remains when the life force has been withdrawn is no longer *annamaya*, but simply *anna,* "food": that is to say, an unstable or perishable organic composite. It is in truth the life force or *prāṇa* that literally "makes" or builds the body out of corporeal substances, as the term "*annamaya-kośa*" affirms.

We need next to recall another fundamental distinction, one that proves decisive for the ontological interpretation of physics: the categorical distinction, namely, between a *corporeal* object X and its associated *physical* object SX.[20] This leaves us with *three* closely associated yet fundamentally different bodies to think about: LX, X, and SX, namely. It is to be noted, first of all, that the neuroscientist is primarily concerned with SX; it is on the physical as distinguished from the corporeal level that synaptic knobs fire, that sodium and potassium gates pump ions to propagate potential spikes along axons, and that long-term memory is explained in Hebbian terms. All this may of course be true; yet it is crucial to realize that *the pranic sheath does not bond with the physical body, but with the corporeal*: not with a body made of molecules, but with a body composed, according to Vedantic doctrine, of five *bhūtas* or "elements"

---

20. The corporeal object is what is known through cognitive sense perception, whereas the physical or "molecular object" is known through the *modus operandi* of physics. See *The Quantum Enigma* (Tacoma, WA: Angelico Press/Sophia Perennis, 2012).

which do not appear on our scientific maps at all, because they pertain to the *essential* as opposed to the *quantitative* order. This is a point of capital importance; to repeat: *one cannot attach a subtle to a molecular body.* And why not? For two reasons: first, because molecules and their aggregates, strictly speaking, constitute a derived or "second" reality, as I have argued more than once,[21] which is tantamount to saying that, from a Vedantic point of view, they do not exist at all; and secondly, because the bonding of which we speak is based upon a kinship of *essences*, which obviously does not extend into a domain from which essences have been excluded by its very definition. What stands at issue in the "bonding" of which we speak is in fact a kinship between the five *tanmātras* or "subtle elements" comprising the *sūkshma-śarīra* and their gross counterparts, the aforesaid *bhūtas*. There are of course correspondences between such traditional or "alchemical" conceptions and contemporary chemical notions, which permit us to speak, for example, of certain chemical substances as igneous or "tejasic,"[22] and so forth; the point to bear in mind, however, is that the respective notions pertain nonetheless to distinct domains which must not be confused.

The language of sheaths or *kośas* suggests that each higher sheath is "within" the sheath that precedes it in the hierarchic order corresponding to our enumeration: the pranic sheath, thus, is within the gross or corporeal, the manasic or mental is within the pranic, and the vijnanic or intellective is within the manasic. But whereas this geometric symbolism is most fitting and virtually indispensable, we need to remind ourselves that the relation between successive *kośas* cannot, strictly speaking, be conceived in spatial terms, inasmuch as the higher sheaths, beginning with the pranic, are not subject to the spatial condition, a bound which in a way defines the corporeal domain. The "inwardness," thus, of the higher sheaths, is not spatial, but ontological, if one may put it so. It should, moreover, be noted that there is yet another spatial symbolism, complementary

---

21. See especially "Eddington and the Primacy of the Corporeal" in *The Wisdom of Ancient Cosmology,* op. cit.

22. The term "*tejas*" refers to the third of the five subtle elements known as *mahābhūtas*, which is "fire."

to the preceding, which conceives of ontological hierarchy in terms of "verticality." The higher $ko\acute{s}as$, thus, like the Biblical *regnum Dei*, are in a way both "above" and "within." It needs further to be pointed out that I have, so far, left out of account the highest sheath of all, namely, the *ānandamaya-kośa,* which is said to constitute the *kāraṇa-śarīra* or causal body to which we have already referred. The latter pertains to the celestial or spiritual realm, what René Guénon terms the domain of formless manifestation, which as such transcends the range of the human individuality. What needs especially to be emphasized is that each sheath depends in its operation upon the next higher, whereas the reverse is by no means the case. The operations of the gross body, thus, depend upon the pranic or vital sheath, which in turns operates in conjunction with the manasic or mental. The latter, which may be termed "lunar" or reflective in relation to the intellective sheath, is in turn dependent upon "the light of the intellect" in the accomplishment of its functions. That "light," moreover, is itself derived from a higher source: from the primary or universal Intellect, namely, termed Mahat or Buddhi, which enters the human individual by way of the *ānandamaya-kośa.* That is indeed "the true Light which lighteth every man that cometh into the world."[23] Remove that Light, and instantly all functions of the human individual—vegetative, sensitive, and intellective—cease. The successive $ko\acute{s}as$, thus, in their concatenation, may be said to constitute a kind of "golden chain" by means of which the gifts of life and intelligence are conveyed to the lower domains, all the way to the corporeal sheath, the "body made of food," where the transmission terminates.

We have so far conceived of the subtle body or *sūkshma-śarīra* in terms of its triadic division into sheaths, without taking cognizance of the fact that it has an organismal unity and a kind of subtle anatomy of its own. Admittedly, this question cannot be approached in terms of spatial conceptions, which, strictly speaking, do not apply

23. John 1:9.

in the subtle realm; yet we can think and speak of that anatomy in terms of analogies to corporeal structures which, in some way, exteriorize or exemplify the former. Now, a main feature of this "subtle anatomy" is given by the system of *nāḍīs*—what Guénon renders by the term "luminous arteries," which one may think of as so many "channels" through which the pranic force can flow—a network in which the principal or central *nāḍī*, named *sushumnā*, plays a definitive role. We may think of the latter as representing the trunk of that "imperishable Ashvattha Tree, with its root above and branches spreading downwards" referred to in the Bhagavad Gita,[24] of which the spinal column cum brain constitutes the outermost exemplification.[25] The fact is that whereas the *kośas* correspond, symbolically speaking, to concentric annular regions, the *nāḍīs* represent radial elements, emanating from a center and tending towards a periphery. It is however to be understood that the center in question is not the true or transcendent Center of the human organism, but constitutes a secondary point of origin, sometimes referred to as the "heart," which we may think of as representing the former on the level of the *sūkshma-śarīra*. And as one might expect, the network of *nāḍīs* is indeed related to the corporeal circulatory system, and likewise to the respiratory, both of which in a way "exteriorize" that nadic system. However, its most intimate connection is clearly with the nervous system on account of the "igneous" nature of neural transmission. One must remember that *prāṇa* is inherently igneous or tejasic; the *prāṇamaya-kośa*, after all, constitutes in a way the mythical "chariot of fire" said to carry or convey the soul. The relation, therefore, of the *prāṇamaya-kośa* and its *nāḍīs* with the nervous system is exceedingly close. We contend, in fact, that there is a special kind of transmission from one to the other, and that *human consciousness, in all of its modes, derives precisely from an interchange between the nadic and the nervous systems.*

The *prāṇamaya-kośa* pervades the entire corporeal body and gives it life. As the vegetative soul, it powers every metabolic and

---

24. Chapter 15, verse 1.

25. The fact that the brain is situated "above" shows that the exemplification or "image" is *inverted*, a point of major significance which we must leave aside.

physiological function from within: every living cell of the body derives its life from that pranic sheath. One must bear in mind that the corporeal body as such is not living, not alive: it is its connection with the *prāṇamaya-kośa* that makes it so. But what about "consciousness"? This too, clearly, derives from the *prāṇamaya-kośa*: it obviously must. But by a different transmission: that is my point. A transmission from where to where? Quite evidently, it must be a transmission between the nadic and the nervous system. There is, first of all, a rudimentary level of consciousness, associated with the autonomous nervous system, which one may designate as "psychosomatic." That consciousness, however, which manifests in sensations such as hunger or pain, is normally eclipsed by higher modes associated with the central nervous system, which one may characterize as "mental." One might say that the two modes of consciousness correspond to different ontological levels: the psychosomatic to the pranic, and the mental to the manasic. Thus, if the former entails a transmission between the corporeal and the pranic sheaths, the latter entails an additional transmission between the pranic and the manasic. Now, what renders this "second transmission" possible, according to Vedantic doctrine, are ten "powers" or faculties termed *indriyas* derived from *manas*, the mental faculty *par excellence*. There are five "sensory" *indriyas*, as one might expect, and five *indriyas* concerned with "motor" functions. But whereas these ten powers are essentially mental, they are relegated to the *prāṇamaya-kośa* on account of their connective function. What, then, is *manas*, the faculty from which the ten *indriyas* are said to descend as from a center? Suffice it to say that it corresponds in a way to our conception of "mind," and can itself be subdivided into three powers answering to the notions of intellect, I-consciousness or *ahankāra*, and the central sense or *sensorium commune*.

It is important to remind ourselves that *manas* interacts neither with the corporeal nor with the molecular body, but with the *living* body, which is not separated from the *prāṇamaya-kośa*: everything hinges upon that bonding, that "fusion" of the two outermost *kośas*. In virtue of this bonding there is an association between nerve channels and corresponding *nāḍīs*, and it is this "nadic connection" that constitutes the vital link in the transmission of sensory information from the brain to *manas* and of motor commands from *manas* to the brain. There is here indeed a "pickup of information" from neuron[26] to *nādī* as well as from *nādī* to neuron: but these are transmissions effected by the very bonding which *defines* the *annamaya-kośa*. Gibson is right: there *is no* "little man" inside the head who "reads" the computer.[27] *There need not be*; for it happens that the *annamaya-kośa* and the *prānamaya-kośa* have been joined so as to constitute one single psychosomatic entity. And let us not fail to note, in reference to the so-called "binding problem," that on this level a first "binding" has already occurred. However, there must also be a higher transmission—from the psychosomatic plane into the *manomaya-kośa*—and this is where the ten *indriyas* enter the picture: we may think of them as "projections of *manas*" into the *praṇamaya-kośa*, and thus into the psychosomatic organism.

Let us now consider the human brain in light of these facts. Neurologically speaking, the brain has sensory inputs and motor outputs, and operates "in between" as a kind of information processing device. In addition to its neurological input and output channels, however, the brain has also "vertical" input and output channels, so to speak, through which it is connected to *manas* or "mind". Now, *it is precisely by way of these vertical connections that the functions which we have previously characterized as non-algorithmic are carried out*, because it is in fact *manas* that does so in conjunction with the brain. On its own, the living brain can only accomplish algorithmic and processing functions: its very composition—the fact that it is "made of neurons"—implies as much. Moreover, these neural operations are associated at most with psychosomatic consciousness, in contrast to the higher non-algorithmic functions, which hinge upon a "seeing" that categorically exceeds the psychosomatic domain.

Now, there are in fact two levels upon which this "seeing" can take place: the manasic and the vijnanic, namely. It needs however to be noted that the manasic is itself in a sense intellective, as is evident from the triadic division of *manas* to which we have referred.

26. Strictly speaking, the "living neuron" pertaining to LX as distinguished from both the "merely corporeal" X and molecular SX!

27. On the Gibsonian theory I refer to Chapter 4.

The notion of "intellect" entails therefore a certain ambiguity even within the sphere of the human individual; and whereas the act of visual perception, for example, is unequivocally manasic, despite its intellective nature,[28] it appears that so-called "intellectual activity" can take place on a vijnanic as well as on a manasic plane. One might speak, in the former case, of "intellect," and in the latter of "reason"; the point, however, is that rationality, too, is inherently intellective. *The fact is that truth can only be grasped through an act of "seeing" which is inherently intellective, no matter on what level it takes place.*

These considerations accord with the central thesis of Roger Penrose, the contention that mathematical discovery and proof do not reduce to algorithmic operations, and thus to brain function, and in a way verify his surmise that "nonlocality" is key to the resolution of the binding problem. So too they accord with William Dembski's thesis to the effect that "intelligent design" cannot be effected by algorithmic means[29]: in the sphere of creative human activity no less than in that of rational thought, an intellective act proves to be pivotal.[30] In a word, all properly *human* actions are intelligent. Not only, therefore, do there exist higher non-algorithmic functions, but these prove in fact to be definitive of the human state. In line with these observations, it can be said that the *normal* level of human consciousness is in fact manasic: man is indeed a "rational animal," that is to say, a *mental* creature.[31] Admittedly, in our integral being we encompass two components or sheaths above the *manomaya-kośa*; yet the fact remains that "normally" we are not

28. See Chapter 4, especially the last two sections.

29. See *The Design Inference* (Cambridge University Press, 1998). For a summary of Dembski's theory I refer to my chapter "Intelligent Design and Vertical Causality" in *The Wisdom of Ancient Cosmology*, op. cit.

30. The intellective basis of all true "art" was well understood in medieval times: as a Scholastic maxim has it: *"Ars sine scientia nihil."*

31. The connection between the English word "man" or the German *"Mensch"* and the Latin *"mens"* may or may not be etymological, but is in any case significant.

conscious on these higher planes. In addition to the manasic level, we tend to be conscious on the psychosomatic as well, but mainly in a peripheral manner; and as everyone knows, to the extent that we become engrossed in authentically *human* activity, the psychosomatic sensations fade into oblivion. One may suppose that even the mathematician operates normally on a manasic level; yet what he does by these means can also be done—and done better!—on a vijnanic plane: this follows from the fact that each (lower) *kośa* operates in conjunction with the higher, from which it derives its principle of operation. I wish now to point out that this explains the experience of Henri Poincaré when he boarded the bus, and of Penrose when he crossed the London street: in either case a "window" to the vijnanic plane was opened for a second or two. It is significant that neither event was prompted by anything taking place at the time, of which the subject was aware; as a rule, the "door" cannot be forced from below. Nor indeed can it be held open; the most that can be done "mentally" is to recall the event and grasp something of its import.

In the waking state, *manas* functions in close relation with the brain: there is actually a division of labor, if you will, between *manas* and brain. Take the case of visual perception: as we have seen, it is the function of the brain "to take the picture apart": to send the retinal input through various filters, as it were, each specific to a certain parameter, be it the orientation of lines, a kind of motion, a color, and so forth. However, it happens *not* to be the function of the brain to *perceive*: the brain is simply not designed "to put it together," and is inherently incapable of such an operation. The brain *separates*, the mind *unites*: that is the plan. It is however to be noted that in the case of visual perception this "binding," which takes place in the *manomaya-kośa*, is not a matter of seeing an image—of "putting the picture together" as Crick has it—but of *perceiving the environment*, which is something else entirely. Admittedly, it is, in a sense, a question of "information pickup"; but what is thus "picked up" is not a mosaic of sense impressions, or of neural firings, but precisely what Gibson terms *invariants*: "forms," namely, in the Scholastic sense.[32]

32. See Chapter 4, pp. 91–93.

It needs to be understood that perception constitutes indeed an act of knowing, in which, "in a certain manner," the subject becomes one with its object, as Aristotle says. Now, it is hardly necessary to point out that such an authentically *intellective* act is incurably non-algorithmic, and exceeds in principle the capacity of brains and Turing machines alike. Of course, *manas* interacts, not just with the visual system, but with numerous other functional units distributed over various regions of the brain; and in each case the FU processes information, of which *manas* avails itself.

This does not, however, mean that *manas* responds to firing patterns involving millions of neurons; for as we have noted, *manas* interacts, not with the corporeal brain—let alone the physical—but with the psychosomatic organism, in which a first binding has already taken place. What *manas* surveys, if one may put it so, is worlds removed from the firing patterns the neuroscientist has his eye upon; in fact, the "information" upon which *manas* draws cannot be conceived in purely *quantitative* terms, but encompasses perforce a qualitative and indeed *essential* content, failing which it would not be in any sense "visible" to *manas*, nor in fact would it *exist*. No wonder neuroscientists have found it hard to explain how a myriad potential spikes can be transformed into perceptions and thoughts, for indeed, no such conversion takes place or *can* take place at all!

There is another major point to be made, which relates to the five so-called *karmendriyas*, the faculties which enable us to perform "voluntary" actions. It is crucial to note that these actions too are perforce non-algorthmic, and do not in fact reduce to brain function: what stands at issue, after all, is what is traditionally termed "freedom of the will." It is clear that neuroscience—or better said, neuroscientism—denies that freedom: "Future generations," writes Rita Carter, "will take it for granted that we are programmable machines, just as we take for granted the fact that the earth is round."[33] Suffice it to note that we have arrived at a very different conclusion: it follows from what has gone before that actions emanating from *manas* involve a mode of causation which is incurably

33. Op. cit., p. 207.

"vertical," and consequently does not reduce to the categories of chance, necessity, or stochastic process.[34]

A comment is in order, finally, regarding the so-called "split mind" of lobotomized subjects: the case of P.S., for instance, who by his left brain wanted to become a draftsman, and by his right a race-car driver. It is to be understood that what has become "split" is not the mind, properly so called, but quite simply the cerebrum. *Manas*, to be sure, can interact with either cerebral hemisphere; and when the two hemispheres have become disconnected through severance of the corpus callosum, these respective interactions can indeed give rise to a different response. Meanwhile neither mind nor consciousness, properly so called, can ever be "split."

In conclusion, let me say a few words, at least, concerning the Vedantic anthropology which we have invoked: what is the basis, let us ask ourselves, upon which that doctrine rests? Obviously it is not "science" in *our* sense of the term. What then could it be? Is it a question of philosophy, a kind of "religious theorizing" perhaps? Fundamentally, I believe, it is a matter of "*seeing*": a discernment that springs from "higher modes of perception." Even as "*the pure in heart shall see God*," so too they shall come to "see" the mysteries of God, including those that underlie what theologians term the Creation. Now, the essential precondition to all such "seeing" is doubtless a radical *metanoia*: a shift of our intellective gaze from the outward or sense-perceived world to the inner, of which most of us have only a kind of "second-hand" or conceptualized notion. What stands at issue is indeed a *self-knowing*, in keeping with the Delphic injunction; and here, again, "doors" need to open, which cannot be pried "from below." The requisite means, therefore, are incurably *initiatic*. The task at issue, in fact, exceeds categorically what the

---

34. On this subject I refer again to my chapter on "Intelligent Design and Vertical Causation" in *The Wisdom of Ancient Cosmology*, op. cit. It turns out, moreover, that vertical causality plays a decisive role in quantum theory, a question on which I refer to Chapter 6 of my monograph, *The Quantum Enigma*, op. cit.

human individual—who is, after all, a creature of the surface-knowing—is able to accomplish on his own. By the same token, moreover, one sees that in this "transcendental" quest the methods of Western science are absolutely of no avail; the very conception of that "inward path" transgresses the horizon of contemporary scientific thought. Our scientists have probed the outer universe—from sub-atomic particles to galaxies supposedly billions of light-years distant—and have now begun to search even for what Crick terms the "soul"; yet, in all these pursuits, they are looking "outwards," towards a periphery which, in the final count, does not exist. As I have argued elsewhere,[35] such knowledge is always mingled with delusion: it is a knowledge that scatters, and in a way perpetuates the Fall. To be sure, a "knowledge" of sorts it is; but not a *jñāna,* not a *gnosis*: not the kind of knowing that can enlighten us regarding God and the *soul.*

35. *Cosmos and Transcendence* (Tacoma, WA: Angelico Press/Sophia Perennis, 2012), pp. 161–66.

6

CAKRA AND PLANET:
O. M. HINZE'S DISCOVERY

IN A SLENDER BOOK entitled *Tantra Vidyā: Archaic Astronomy and Tantra Yoga*,[1] Oscar Marcel Hinze reports on a scientific discovery, the implications of which are epochal. The work consists of three essays, previously published in the German-speaking domain, which deal respectively with archaic astronomy, Tantric Yoga, and surprisingly enough, with the teachings of Parmenides. Contending that these seemingly disparate topics are intimately linked, Hinze proceeds to demonstrate that connection by eliciting concordances so striking and so precise as to dispel all reasonable doubt. It emerges, first of all, that the Gestalt aspects of planetary astronomy stand to the Tantric *cakra* anatomy as macrocosmic and microcosmic manifestations, respectively, of one and the same paradigmatic structure. A hitherto unsurmised isomorphism between the planetary system and the subtle anatomy of man has thus come to light, which is to say that the structural identity of macro and microcosm, as traditionally conceived, has now been corroborated on sober scientific ground. Such is the burden of the first two essays; and the third proves to be no less momentous. First published in 1971, it establishes Oscar Marcel Hinze as one of the earliest authors to have rediscovered the "true face" of Parmenides, a face hidden for well over two thousand years: in place of the legendary "logician" who supposedly propounded a world-denying monism, he reveals to us an adept of Kundalini Yoga, speaking from the plane of the *ājñācakra*. And again Hinze makes his point, not by way of vague

1. New Delhi: Motilal Banarsidass, 2002.

131

speculations, but on the strength of precise concordances too cogent, by far, to be dismissed as mere "coincidence." In a word, Hinze's ground-breaking treatise constitutes a pivotal contribution to the ongoing rediscovery of the authentic *cosmologia perennis*.

Part One, as we have said, deals with archaic astronomy, a science based upon direct visual observation of the night sky. One needs however to understand that there are different grades and modes of observation, and that the human faculties or powers of perception were incomparably greater in archaic times than they are today. In the spirit of gestalt psychology, Hinze maintains that what is perceived primarily is the whole as distinguished from its parts; yet he contends that in course of time, be it in the development of the individual or of the race, the balance shifts gradually from the whole to the parts. Based upon psychological as well as anthropological data, Hinze maintains that the child, no less than archaic man, perceives Gestalt first and foremost, whereas we, adults of the present day, perceive mainly an aggregate of parts. In a word, *human perception tends to disintegrate.* It is easy to understand, moreover, that the rise to dominance of modern science has significantly exacerbated this universal trend; in the wake of what historians term the Enlightenment it appears that our capacity to discern the Gestalt of natural phenomena has been vastly diminished. Our reductionist philosophy, meanwhile, bestows ontological primacy upon the parts, and ultimately upon the quantitative residue that remains when all wholes, and thus all essence or being, have been evacuated from the world.[2]

Clearly, these observations open new vistas in our comprehension of archaic astronomy; as the author observes: "The planets in ancient times were not independent pieces of matter somewhere in empty space, but organic parts of the archaic sky which maintain their qualities and importance by virtue of their respective positions

---

2. See Chapter 2 for an in-depth discussion of this question.

within the whole." $(8)^3$ The crucial point to be noted is that these "qualities" do no exist from a reductionist point of view: they do not pertain to the planets conceived as "independent pieces of matter somewhere in empty space." Yet the scientist of today is thoroughly mistaken when he concludes that the qualities in question are consequently imaginary or unreal. The disappearance of the qualities, Hinze argues, so far from being mandated by scientific enlightenment, is caused primarily by the previously mentioned decline in our ability to perceive Gestalt. Add to this diminution the fact that modern astronomy is based upon artificial means of observation designed to detect and measure *quantitative* parameters—and one sees why the very content of archaic astronomy has disappeared from scientific view.

What proves crucial for archaic astronomy is what Hinze terms "successive Gestalt," that is to say, Gestalt given in the successive positions of a body over some span of time. It should be noted that this kind of Gestalt is still perceivable for us, provided the corresponding time span is sufficiently short. The most obvious examples pertain to the auditory domain: our ability, for instance, to "hear" melodies and words. We can also, however, perceive successive Gestalt visually, as in the case of a dance. Hinze concludes that "There are, therefore, perceptions which, without losing their unity or clarity, fill out a certain span of time and can have a temporal content of this span as object."⁴ (12) He speaks in this connection of a *"presence-time,"* and enunciates a law of decline, both ontogenetic and phylogenetic, with reference to this parameter. It appears—surprising as this may seem—that archaic man disposed over presence-times large enough to bring the successive Gestalt of planetary motions into the range of visual perception. There is, in fact, reason to believe that the hard and fast distinction between memory and perception, which we are accustomed to draw, was largely transcended in archaic times:

3. Numbers in parentheses designate page numbers in Hinze's book, op. cit.

4. The reader will note that this agrees with James Gibson's fundamental claim to the effect that we perceive movement and events without the intervention of memory. See Chapter 4, pp. 80–83.

We must think that the ancient observers of the sky must have been equipped with an extraordinarily vital perceiving memory, with the capacity to view together in a present unity phenomena which, today for us, lie temporally too far apart to be still perceived as belonging together. (21)

So too there is reason to believe that the categorical separation between the visual, the auditory, and other sensory domains was likewise transcended. It should be mentioned that there exists a considerable body of evidence in support of this contention; in an experiment involving persons under the influence of Mescaline, for instance, the subject had this to say: "I felt, saw, tasted, and smelt sound. I was myself sound."[5] In short, based upon various kinds of evidence, Hinze formulates another genetic law: "The further one recedes back in the development, the more the individual areas of sense, which in the culturally formed man (*"Kulturmensch"*) of today are clearly differentiated from one another, are found still united."

It has become evident that archaic man had access to realms of sensory experience which for us are closed. It is not however a matter, finally, of sensory domains, but of *meaning*: of *access to archetypes.* It is a question of reading the Book of Nature, of perceiving "the invisible things of God in the things that are made." In the case of astronomy, of course, what is to be "read" is principally the night sky: the "lights in the firmament of heaven," which were given to us not only "for seasons, and for days, and years," but above all "for signs." When Hinze speaks of "Gestalt astronomy," we must remember that "Gestalt" signifies incomparably more than a mere visual form, figure, or pattern: what ultimately stands at issue is the miracle of *semanticity*, of a sign that presentifies a transcendent referent. As Hinze explains: "The archaic priests who watched the heavens comprehended, at the highest stage of their graphological interpretation of the sky, the stars and their movements as cosmic symbols which, when thus deciphered, provide explanation concerning the most essential questions of human life." (23) One cannot

5. Heinz Werner, *Comparative Psychology of Mental Development* (NY, 1948), p. 68. Quoted by Hinze in op. cit.

but agree that this ancient science was indeed a *"symbolical Gestalt-astronomy."*[6]

In the second part of his treatise the author introduces us to the fundamental conceptions of *tantra vidyā*, beginning with the seven principal "centers" in man, symbolically designated as *cakras* ("wheels" or "circles") or *padmas* ("lotus flowers"). As one knows, these centers are situated along an axis corresponding to the spinal column, and range from the *mūlādhāra-cakra* near its base to the *sahasrāra* at the crown of the head. Each center is characterized by an integer, which we may think of as the number of "petals" of the corresponding *padma* (or of "spokes" according to the *cakra* symbolism). The resulting sequence, taken in ascending order (from *mūlādhāra* to *sahasrāra*), is 4, 6, 10, 12, 16, 2, 1000. Following Hinze, we will sometimes find it convenient to designate a center by its associated petal-number; thus (4) will designate the *mūlādhāra*, and so forth. It is to be noted that the sum of the first six numbers in the sequence is 50, which is also the number of letters in the (cultic) Devanāgarī alphabet. It should be emphasized that this connection between *cakras* and sound or speech proves to be basic: Tantric tradition conceives the creation, both in its macrocosmic and microcosmic aspect, as an effect of *śabda-brahman*, which literally means "sound Brahman," that is to say, God manifesting as sound or speech, a notion which recalls what Christianity knows as the Word of God.[7] Here too can it be said that *"In the beginning was the Word,"* though to be sure, Tantric tradition understands this in its own way. From that *śabda-brahman* or Word, in any case, have sprung, on the one hand, the worlds or *lokas,* and on the other, the human microcosm, beginning with the hierarchy of *cakras.* Leaving aside the *sahasrāra*, which is represented symbolically as a 1000-

---

6. Or as one may also say: an *astrology* in the true sense.

7. It is to be noted that though the *śabda-brahman* does in a way correspond to the Logos or Word, *tantra vidyā* does not, most assuredly, conceive of that *śabda-brahman* in Trinitarian terms.

petalled lotus (and is never referred to as a *cakra*), every *cakra* is in fact associated with a *bījamantra* or "seed sound," which Hinze refers to as its Central Sound. From this Central Sound arise various differentiated sounds, which are precisely the sounds represented by the letters of the Devanāgari alphabet, and which correspond to the "petals" of the associated *padma* or lotus. One sees that the connection between *padmas* and letters of the alphabet stipulated in *tantra vidyā* is by no means adventitious, to say the least. We should mention next that there is a connection between the first five *cakras* and the five classical elements; to be precise, (4) corresponds to earth, (6) to water, (10) to fire, (12) to air, and (16) to the ether. According to Tantric doctrine, each element springs from the corresponding *bījamantra* as what might be termed its elemental manifestation. If we now consider the distribution of the *cakras* within the human body, we find that the first four are situated in the trunk, the fifth in the throat, and the remaining two in the head. Again leaving the *sahasrāra* out of account (in view of its "transcendent" nature), we recover thus the traditional division of the *tribhuvana* or triple world, consisting of the "earthly" realm, made up of the four "differentiated" elements and represented microcosmically by the human torso, the "intermediary" realm, associated with the fifth element or *quinta essentia* (which contains the four lower elements synthetically) and is represented in the human body by the throat or neck, plus the third or "celestial" realm, corresponding to the sixth center, the *ājñā-cakra*, represented in the body by the head and traditionally depicted as a "third eye" at the center of the forehead. In contrast to the first five *cakras*, the *ājñā-cakra* is not associated with any element, but corresponds to what may be termed the spiritual nature of man (the *antaḥkaraṇa* or "inner instrument," consisting of *manas*, *buddhi*, and *ahaṁkāra*). We cannot, of course, enter here into a detailed discussion of these matters; suffice it to reiterate that the lower six *cakras*, grouped according to their respective positions within the torso, neck and head, do correspond, quite visibly, to the divisions of the Vedic *tribhuvana*. With the exception of the *yantras* or geometric figures, the remaining symbolic elements entering into the traditional description of the *cakras*, such as the various "divinities" or "embodiments of *Śakti*"[8] revealed at these loci, or again, the symbolic animals constituting their *vāhana* or "vehicles," are of secondary interest, given the focus of our treatise. Hinze himself touches upon these matters but lightly, in keeping with the fact that his primary concern is the discovery of a concordance between the Tantric symbolism and archaic astronomy; and clearly, what lends itself most readily to that end are the numerical and geometric aspects of the Tantric description.

The author approaches the problem by way of texts outside the Tantric domain, beginning with *Saundaryalahari*, a Sanskrit poem attributed to Shankarāchārya. Here, in this little known text, the seven principal "centers" are clearly mentioned, but only in their macrocosmic manifestations as "circles or spheres in the universe." As Hinze explains: "The universe is understood as the body of the divine world-mother *Mahā-Devi*; it has developed out of a preworld state (*sahasrāra*) in six steps, of which the first corresponds to *ājñā-cakra*." (39) One is of course reminded of our "six Days of creation." Now, considering the extraordinary interest in "the lights of the firmament," and in particular, in the seven planetary bodies, evinced by archaic man, one cannot but ask whether the seven cosmic regions of the *Saundaryalahari* do not in fact correspond to the seven classical planets. "This question," Hinze concludes, "cannot be answered with certainty on the basis of the *Saundaryalahari* alone." (40) For additional evidence he turns to Mithraism; as Franz Cumont observes in *The Mysteries of Mithra*: "The seven steps of Initiation which the mystic has to undergo in order to attain perfect wisdom and purity correspond in this cult to the spheres of the seven planets." But still we lack the key, the Rosetta stone if you will; and this is what Hinze eventually discovered—surprisingly enough—in a work by Johann Georg Gichtel, a disciple of Jakob Boehme. It is given in an illustration found in his *Theosophia Practica*, first published in 1696, depicting the seven centers within the human body, labeled with the corresponding planetary signs. The order of the planets is that of Ptolemy: Moon, Mercury, Venus, Sun,

8. The term *Śakti* refers to the "power aspect" of God, or better said, to the feminine component of the divine bi-unity (designated by the term *Śiva-Śakti*).

Mars, Jupiter, and Saturn. How, then, did Gichtel arrive at this correspondence? On the title page of his book the German theosopher himself provides the essential clue: "A short explanation of the three principles of the three worlds in man, represented in clear illustrations which show how and where they have their respective centers, *according to what the author in his divine contemplations has found in himself and what he has felt, experienced and perceived.*"

With this correspondence between *cakras* and planets in place, Hinze turns to the accomplishment of his principal task, which is *to account for the petal-numbers in astronomical terms.*[9] Here the real work begins. The first thing that needs to be done is to distinguish between "period-numbers," which depend on a particular unit of time (such as the year, the month, or the day) and "gestalt-numbers," which are descriptive of a geometric figure swept out by planetary bodies, and having done so, to recognize that what counts microcosmically are in fact the latter. Hinze's problem, now, is to show that the Moon has the gestalt-number 4, Mercury the gestalt-number 6, and so forth up the line; the case of Saturn (corresponding to the 1000-petalled lotus named *sahasrāra*) is of course unique, and requires special considerations appropriate to the transcendent nature of that supreme Center. Inasmuch as the resultant investigations are perforce technical, I will not provide a complete explanation, but will attempt simply to convey the idea of what is involved.[10] To begin with, I will note that the gestalt-number of the Moon is indeed 4 on account of the rectangular configuration defined by its four recognized phases. Since one of these is invisible, the 4 breaks up into 1 + 3, which constitutes what Hinze calls a secondary correspondence relating to the letters of the Devanāgarī alphabet associated with the given lotus: for it happens that the four letters corresponding to the *mūlādhāra-cakra* consist of 1 semivowel and 3 sibilants. Getting back to the Moon, it is of interest to note that this planet forms also a second figure, which was likewise well known in ancient times: a triangle, namely; and so, too, one

---

9. It may be safe to say that no one prior to Hinze has ever accomplished that feat, or perhaps even conceived of that possibility.

10. The full exposition is to be found in *Tantra Vidyā,* op. cit., pp. 42–75.

finds, in traditional iconographic representations of the *mūlādhāra*, a triangle inscribed within a square.

Turning next to the planet Mercury, which according to Gichtel corresponds to the second *cakra* (having a petal-number 6), it can be shown that its conjunctions with the Sun give rise to a hexagram, which in fact consists of two superposed triangles. This not only gives a gestalt-number 6, but also defines a secondary correspondence with the associated letters, which consist of 3 labials and 3 semi-vowels. Next in line is Venus, a planet which in association with the Zodiac gives rise to a figure consisting of two pentagrams; and here again one has a primary correspondence with the associated *cakra* (whose petal-number is 10), plus what Hinze terms a "partial" secondary correspondence with the letters in question, which consist of 5 dentals, 3 cerebrals and 2 labials. What stands at issue here, in Hinze's view, is that a third order correspondence enters the picture, involving a further division of 5 into 3 + 2. With this understanding, it can be said that "as a matter of fact, all lotus-flowers exhibit a correspondence of second order to the planets belonging to them." Turning now to the Sun, one finds that it has a gestalt-number arising from its conjunctions and oppositions with the Moon. The key is to consider the two special solar eclipses, one in the ascending mode of the Moon (known as *rahu* in Vedic India), and one in its descending mode (known as *ketu*). As Hinze goes on to point out: "The position of these solar eclipses in the Zodiac is such that they lie opposite to each other and on both sides 5 conjunctions take place. The astronomical structure of the Sun-Moon meetings is thus actually organized as 5 + 5 + 2 = 12. The distribution of letters in the 12-petalled lotus [into 5 gutturals, 5 palatals, and 2 cerebrals] constitutes the exact expression of this astronomical structure." (64)

This brings us to the planet Mars, which as usual proves to be of particular interest as well as difficulty. One may recall that it was the obstreperous behavior of this planet that prompted Kepler to break at last with the customary assumptions and inaugurate the era of modern planetary astronomy with a treatise *On the Motion of Mars*, published in 1609; and now, again, our search for the gestalt-number of Mars begins with an enigma. It is well known that Ptolemy associates with Mars the period-number 15; but what is it, precisely, that takes place supposedly within that period? A footnote to the Greek-English edition of Ptolemy, published in 1956, informs us that this is "a mystery"! Now, it appears that Hinze has resolved this long-standing enigma through a careful examination of what are termed the "loops" of Mars, of which there are 8, with a total period, yes, of 15 years. These loops are connected with the retrograde movements of Mars, and occur near points at which the planet attains its greatest proximity to the Earth. The loops are therefore clearly visible and are marked by an exceptional brightness. They begin with a slowing of the eastward movement which terminates in a point of reversal (what Hinze refers to as the end of Phase 1), which is followed by a second point of reversal (marking the end of Phase 2), after which the planet resumes its normal eastward course. It is this bipartition of each of the 8 Martian loops that gives rise to the gestalt-number 16, which is indeed the petal-number of the corresponding *cakra*: the fifth, that is, named *viśuddha*, which is associated with the element *ākāśa* and is located in the throat. It remains to be noted that the 16 letters associated with this *cakra* consist precisely of the 16 Sanskrit vowels, which are separated into 8 pairs, consisting of their long and short forms. *I find it utterly fascinating that the phonetic division into long and short forms of the 8 primary vowels should be reflected macrocosmically in the bipartition of 8 Martian loops, a drama which plays itself out in the sky every 15 years!*

I will mention, in passing, that there exists a second approach to the gestalt-number of Mars (involving "two pairs of 8 Venus-Moon meetings"), which from an astrological point of view could be described as complementary to the first. This complementarity corresponds moreover to a complementarity between the first and the fifth *cakras*, recognized in the Tantric tradition, which moreover accords with the fact that these respective centers have the same symbolic animal (the elephant, namely) as their *vāha*. Again it would take us too far afield to follow Hinze in his elucidation of these various connections; suffice it to cite his concluding remarks: "When, therefore, Gichtel localizes Mars in the larynx while the customary astrological tradition puts here Taurus (and therefore the planets Venus and Moon), and when, furthermore, the Indian tradition sets in this place a 16-petalled flower, we have to do in all these cases with an exact representation of Gestalt-astronomy." By way of additional corroboration, our polymath author presents the figure of a bull's head carved in silver, found at Mykene, with a 16-petalled rosette on its forehead; and he displays the title-page of a Dutch book concerning the South American Indians of Surinam, depicting a warrior named "Kainema" (meaning "blood-feud"), marked with a star-like figure of 16 rays centered at the throat. After relating how the historical Kainema was charged with the duty of avenging the violent death of his father, Hinze concludes:

> Now Gichtel localizes the planet Mars in the same place of the body where also Kainema is strikingly marked. But as is well known, Mars also points to aggression, power and violence. Finally, when one reads the text on the 16-petalled lotus in the *Sat-cakra-nirūpana* [the primary text on Kundalini Yoga] where it is reported that the yogi who rules this center will be able to move all the three worlds "in his anger," one sees how also here is pointed out the character of violence and power of this center. (72)

Above the Martian center in the throat stands the *ājñā-cakra*, traditionally depicted between the eyebrows, whose lotus has just 2 petals, corresponding to the Sun and Moon. As noted before, this center pertains to the spiritual realm, the third and highest "world" of the *tribhuvana*. It is here, in connection with this supreme *loka* properly so called, that Hinze unfolds his most beautiful and penetrating elucidations, which however we will not attempt to set forth here. Suffice it to point out that the *ājñā-cakra* represents neither the Sun nor the Moon, but their place of meeting, which constitutes a kind of "celestial Heart." To be precise, the *ājñā-cakra* is the center where the *nādīs*[11] named *pingalā* and *idā* (corresponding to Sun and Moon, respectively) meet, a juncture, let us add, depicted in the familiar figure of the Hermetic *caduceus*. As is only to be expected,

---

11. The term refers to "nerves," or more precisely, "channels" pertaining to the subtle anatomy. See Chapter 5, pp. 122–126.

the combined astrological and alchemical symbolism relating to the *cakra* in question—ruled by the "royal planet" Jupiter!—is rich enough to fill a treatise of its own. It should be mentioned that the author's inquiry into this *cakra* leads him to consider a so-called secondary *cakra* named *dvādaśārṇa*, with a petal-number 12, situated between the *ājñā-cakra* and the *sahasrāra*. The fact that *dvādaśārṇa* is associated with 2 letters, each occurring 6 times, connects it structurally with the *ājñā-cakra*, and it is by way of this connection, precisely, that the correspondence between the sixth primary *cakra* and the planet Jupiter comes to light. I would like to add that the secondary centers are of interest in other respects as well; one needs to understand that despite the primacy or dominance of the seven "classical" centers, the number of *cakras* is said to be "*ananta*," that is, "unbounded," or as we also say, "infinite." It stands to reason, therefore, that an astrology based upon seven planets cannot be fully comprehensive, and that in principle secondary "planets" need also to be considered. One sees that the discovery of additional planetary bodies, beginning with Uranus and Pluto, does not by any means conflict with the principles of authentic astrology.

We have so far left out of consideration the highest center, the *sahasrāra*, symbolized by a 1000-petalled lotus. Speaking in mathematical terms, one could say that it represents, not the last term of a series, but rather its limit; as Hinze observes: "The 1000-petalled lotus-flower is already the superhuman in man." The corresponding "*loka*,"—which is not, strictly speaking, a *loka* or "world" at all— does not enter into relation with anything else, and can be spoken of only in apophatic terms. It is to be known in the state of *nirvikalpa samādhi*, which in fact is said to result from an "awakening" of the *sahasrāra*. The petal-number itself apprises us of the fact that all things found in the six *lokas* of the *tribhuvana* are contained preeminently in that transcendent state: such is the symbolic reading of the number 1000. The adjunct symbolism of letters, moreover, accords with this fact; for it is said that each letter of the Devanāgarī alphabet occurs 20 times in the petals of the *sahasrāra*. There are, of course, astrological reasons why the planet Saturn is associated with this last and highest center; it should however be apparent that this correspondence can have little to do any more with gestalt-numbers.

Getting back to the array of concordances between the Tantric *cakra* anatomy and the facts of Gestalt-astronomy, I would like to point out that these remarkable discoveries actually exonerate the much-ridiculed tenet of "geocentrism"; for it happens that the astronomy which yields the concordances in question is incurably geocentric. Take the loops of Mars, for example: these simply do not exist from a heliocentric perspective, nor from the purview of contemporary cosmology, founded as it is upon the so-called Copernican principle.[12] The fact, therefore, that a congruence between the human microcosm and the planetary macrocosm comes to light precisely from a geocentrist point of view bestows not only legitimacy, but indeed a certain primacy upon geocentric cosmology. One sees that the so-called Copernican Revolution, which has been depicted in our schools and universities as a victory of science over superstition, is in reality the fateful step that has closed the door to any higher understanding of man and his destiny.

There is, however, a second major point to be made: it turns out that Hinze's discovery disqualifies contemporary scientific theories concerning the origin of our planetary system. It does so, moreover, at one stroke, and with exemplary rigor, due to the fact that the concordances in question translate into so-called "complex specified information": for one knows today, on the strength of a mathematical theorem, that no natural process, be it deterministic, random, or stochastic, can generate CSI.[13] I am referring, of course, to what has come to be known as the theory of "intelligent design," a subject which of late has received considerable attention in scientific circles and in the media. Unfortunately, however, the theory has almost invariably and indeed tendentiously been misconstrued as "creationist," when it happens to be actually the only bit of "hard" science bearing directly and decisively upon the issue in question. So far from being based upon religious faith or Biblical convictions, ID

12. See Chapter 1, pp. 14–17.

13. See William A. Dembski, *The Design Inference* (Cambridge University Press, 1998). For a readable summary of ID theory I refer to Chapter 10 of my book, *The Wisdom of Ancient Cosmology* (Oakton, VA: Foundation for Traditional Studies, 2003).

theory rests upon a theorem or law as solid as the Second Law of thermodynamics, to which in fact it is closely related. To set the record straight: it is the Darwinists, and not their ID opponents, who are violating the norms of scientific argument. Getting back to the planetary system, it is now demonstrable that no "naturalistic" explanation of its origin—no explanation based upon deterministic, random, or stochastic processes, to be exact—can account for the concordances which Hinze has brought to light. It is with the planetary system as it is with the genome: the fact that these structures "carry CSI" brings into play perforce a notion of "intelligent design," or "vertical causation."[14]

In the third part of his book Hinze reflects upon the teachings of Parmenides, which he approaches from an inherently yogic or "initiatic" point of view. He is interested not only in the doctrine, but especially in its author, in the genre, one might say, of the man; and it happens that there are major clues that prove to be enlightening in that regard. The figure of Parmenides that emerges from the resultant elucidations differs sharply from that of the "world-denying logician" to be found in textbooks of philosophy: in place of a mere "thinker" Hinze reveals to us the lineaments of an adept, someone who has broken through to a higher mode of knowing. So too Hinze gives us to understand that what Parmenides teaches is not the reputed "monism" discussed in departments of philosophy, but something reminiscent, at least, of *advaita*, the veritable "non-dualism" as is to be found, for instance, in the Upanishads.

It appears however that the stereotype of the "quixotic logician," so far from constituting a modern invention, goes back a long way: to the days of Aristotle in fact, who regarded the Parmenidean "monism" as something "close to madness." Yet that "madness" has proved to be singularly seminal: for at least a century following the demise of

14. Regarding vertical causation, see my monograph *The Quantum Enigma* (Tacoma, WA: Angelico Press/Sophia Perennis, 2012), chap. 6.

the Master, it was his doctrine that exercised the leading thinkers of the time, to the point that the so-called "miracle of Greek philosophy" may actually be seen as a response to his teaching. But what was that teaching: was it a spurious monism, or authentic *advaita*? It is hard to say; what is clear, in any case, is that by the time we reach Aristotle, all that remained was a doctrine "close to madness."

I find it noteworthy that Parmenides stands not only at the beginning of what is often termed Western culture, but that he emerges again near the end: for he has in fact been "rediscovered" in our day, and the authentic sense of his teachings continues to be uncovered by scholars, howbeit outside the academic mainstream. A case in point, evidently, is O.M. Hinze, whose article on the subject (first published in 1971) constitutes one of the early studies belonging to this new genre. Sporadically, and certainly without fanfare or approbation from on high, a handful of scholars, imbued with a certain knowledge of Eastern traditions, have come forward to re-examine the legacy of the Presocratics, and in so doing have brought to light truths that had long been buried under the sands of time; as Peter Kingsley remarks in the opening sentence of his own magisterial treatise: "I had better write these things down before they are lost for another two thousand years."[15]

The teaching of Parmenides, as one knows, is given in a single didactic poem which has come down to us in fragments, transmitted by various authors of antiquity. It begins with the description of a Journey into "the Mansions of Night," the realm of the dead ruled by the Goddess Persephone. Is it not amazing that this "detail" should for so long have escaped serious attention at the hands of leading exegetes? Would not an educated Hindu, for example, be reminded instantly of Nachiketā, who likewise traveled to the underworld, as one reads in the *Katha Upanishad*, in search of truth? The Journey undertaken by Parmenides is described in astonishing detail, every facet of which has presumably its significance. There is

---

15. *Reality* (Inverness, CA: The Golden Sufi Center, 2003). Since the publication of his first book (*Ancient Philosophy, Mystery and Magic*, Oxford University Press, 1995), Kinglsey ranks as one of the leading authorities on Presocratic philosophy.

reference, thus, to a "chariot" drawn by "horses" (by "mares" to be exact), guided by "maidens" said to be "daughters of the Sun"; there is mention of an "axle" and of "hubs" and "wheels," and an allusion to "gates" that open and shut. And most importantly, it is only at the end of the Journey that the now famous Doctrine is communicated to Parmenides by the Goddess herself: "I will do the talking," she says to him, "and it is up to you to carry away my words once you have heard them." As Hinze and Kingsley both point out, so far from a logician pondering syllogisms, Parmenides is basically a prophet: a messenger, that is, from a realm beyond this world. But apparently this "detail" has generally been missed by historians and philosophers alike; and this is what in a way predetermines the end result of their exegesis: the premises more or less entail the conclusion. Having ignored or "explained away" the Journey, and relegated the figure of the Goddess into whose realm Parmenides was conveyed to the status of a literary device, is it any wonder that our erudite "experts" have roundly misunderstood the Doctrine itself?

What further confounds the exegetes is the fact that the Goddess professes, not one, but *two* doctrines, which moreover appear to be logically incompatible. The usual response to this impasse has been to demote and in effect eliminate Part Two of the didactic poem, a strategy which in fact traces back to Aristotle, who thought that in Part Two Parmenides is simply recounting the views of his predecessors with the intention of rejecting them. Yet it turns out that the two parts belong together, that in fact they complement and complete each other; Hinze has undoubtedly hit the nail on the head when he writes:

> The subdivision of the doctrinal poem into two sections, of which the first deals with "Being" and the absolute truth, and the second with "appearance" and "the meanings of mortals," has its exact correspondence in India with the doctrine of the two kinds or levels of knowledge, of which one is called the "higher" and the other the "lower" knowledge. (84)

What stands at issue here is the Vedic distinction between *para-vidyā* or "supreme knowing" based upon *anubhava*, the unmediated perception of the highest reality,[16] and *aparavidyā*, a lower or "non-supreme" knowing pertaining to what might be termed the realm of "appearances" in the widest conceivable sense. We need to understand, in the first place, that the latter category includes all that we rightly regard as "knowledge," be it of cosmic or of supra-cosmic realities. Despite its "lesser" status, moreover, this *aparavidyā* is not to be despised, neglected, or discarded so long as the *paravidyā* has not been attained; as we read in Mundaka Upanishad[17]: "*Dve vidyā veditavye*" ("Two kinds of knowledge are to be known"). The point is that the Goddess teaches the same: she too does not confine her discourse to "Being and the absolute truth," but goes on to deliver the *aparavidyā* as well. She does so, however, with a warning: from here on, she tells us, her words are "deceptive." And this too, let us note, accords with the Vedantic position. Shankarāchārya, in fact, puts the matter in even stronger terms: in his commentary on the aforesaid Upanishadic verse he refers to the *aparavidyā* as *avidyā* ("ignorance"), a designation which seems to contradict the notion that it is yet a *vidyā*, "lower" (*apara*) though it be. One might say that whereas the Goddess refers to the lower knowing as "deceptive," Shankarāchāryā calls it "deceived." Be that as it may, it should in any case be clear that the teaching transmitted by Parmenides on the subject of "Being and absolute truth" is bound to remain incomprehensible on the plane of *aparavidyā*; as Sri Ramakrishna once put it: "One cannot pour four seers of milk into a three-seer pot."[18] But this, quite apparently, is what most so-called experts on "Presocratic philosophy" have failed to understand.

16. We shall return to the subject of "unmediated perception" in Chapter 8 from a Christian point of view, based upon the teachings of Meister Eckhart.

17. I.i.4.

18. The numbers are significant: as there are three principal "worlds" in the Vedic enumeration (the so-called *tribhuvana*), so there are three associated degrees of knowing (corresponding to the waking state, the dream state, and *sushupti*, the state of dreamless sleep). The "four seers of milk" correspond evidently to the state known as *turīya*, which literally means "the fourth." The Master is saying that what is realized in supreme gnosis is not comprehensible to any lower mode of knowing: "even as four seers of milk cannot be poured into a three-seer pot."

Hinze approaches the subject of the two-fold Doctrine by reflecting on what has sometimes been referred to as the Divine Bi-Unity. To put it in Vedantic terms: the production of the world is to be ascribed, not to *Śiva*, but to *Śakti*. As Hinze explains: "She is the creative Force of God (*Śiva*) and represents His female aspect." (95) The term "aspect" is crucial, for we must not lose sight of the fact that "*Śiva* and *Śakti* are fundamentally one," as Hinze goes on to say. We have here an authentic Mystery, not unlike that of the Trinity, which as one knows, stands at the heart of the Christian teaching. Now, as Hinze points out, it is *Śakti* that gives rise to the cosmic manifestation; but this "feminine aspect of God" has itself two aspects or faces, designated as *Māyā-Śakti* and *Vidyā-Śakti*, which moreover correspond, quite precisely, to Aphrodite and Persephone, respectively, in the Greek tradition. The first-named aspect may be characterized as a power of veiling, which seemingly accomplishes "an act of self-limitation or even self-negation of God" (96); and it is by virtue of this, to us inscrutable Power, that *avidyā*—a kind of universal delusion— enters into the very fabric of cosmic existence. It is essential to understand that the delusion or "ignorance" of which the sages speak is not of our making, but something in which we share by virtue of being what we are, much as we share in what Christianity terms Original Sin. It cannot, therefore, be overcome by the human individual "on his own": *what Śakti binds, Śakti alone can set free.* And this is where *Vidyā-Śakti* comes into play: whereas Aphrodite binds, deludes, and finally kills, Persephone liberates, enlightens, and gives life. But again, let us bear in mind that the two are not separate and opposing Powers, but complementary aspects of a single *Śakti*, which is none other than "the female aspect of God." This calls to mind the spectacle of the young Ramakrishna offering worship to the Divine Mother before the blood-stained image of Kālī, a practice that may strike the Western observer as quite incongruous; and yet, do not we Christians pray daily: "And lead us not into temptation…"? And did not the Garden of Eden already harbor a snake?

Given the stated ambivalence of *Śakt*, it is hardly surprising that there are in fact two principal ways of viewing the cosmos: the first, termed *Vivarta-vāda*, perceives the universe as illusory or "dreamlike," whereas the second, termed *Pariṇāma-vāda*, speaks not of "illusion," but of *flux*, of actual "genesis" or "becoming," a position which in fact is characteristic of Tantrism.[19] Where then, let us ask, does the Goddess stand on the issue in the "cosmological" half of her discourse? According to Hinze, she stands on the side of *Vivarta-vāda*, the side of "illusion" as one might say. Now, granting that this is arguably the case, I find it nonetheless questionable: after all, it may reasonably be surmised that the Doctrine of a Goddess must in fact transcend the confines of a particular *darśana*[20] or "angle of vision"! In the final count there is actually no contradiction between *Vivarta-vāda* and *Pariṇāma-vāda*: "they are only two different ways of looking at the same thing" as Hinze himself points out.

In the first part of her discourse, the Goddess seems to stipulate a dichotomy between Being and Non-Being, and having done so, banish the latter by way of the twin recognitions that "the Being is" and "the Non-Being is not." To the logician or rationalist philosopher, this may imply that Being alone remains—"in splendid isolation" if you will—and that, consequently, there can be in truth no generation or dissolution, no change or motion, no divisions or bounds. Yet Non-Being refuses to be exorcised: in the final count there can be no cosmos, no creation, no universe without Non-Being. And this holds true whether we look at the cosmos from the standpoint of *Vivarta* or *Pariṇāma-vāda*; in either case, Non-Being enters the picture. From the cosmos in its entirety to the least of its parts, everywhere we encounter both Being and Non-Being, as the Goddess herself affirms when she declares, in Part Two of her discourse, that "Everything is at the same time full of Light and lightless Night." Nor is this strange or incongruous if only one recalls that the cosmos, after all, is indeed the manifestation of the Divine Bi-Unity itself; as we read in a Tantric text: "Whatever comes into the world consists of *Śiva* and *Śakti*." It emerges that despite differ-

19. It is to be understood that these alternatives are not mutually exclusive but complementary.

20. Hindu doctrine is traditionally divided into six *darśanas*, sometimes referred to in the West as the "six systems of philosophy." This is however misleading: a *darśana* is a perspective determined by a point of view. There are six basic *darśanas*, just as there are six directions in space. Thus, where the West senses contradiction, the East perceives *complementarity*.

ences of terminology, of "cultural coloration," and perhaps even of *darśana*, the cosmological teaching of Parmenides accords with the Tantric, a conclusion which, as Hinze informs us, "can be confirmed to a still greater extent by a thorough analysis of some other Fragments."

It remains now to consider the introductory portion of the didactic poem, which deals, not with the Doctrine, but with the Journey. It is here that Hinze discovers an impressive array of parallels with *Kundalinī Yoga*, beginning in fact with the very first word of the Greek text: *hippoi*. It is not strange, of course, that a chariot should be drawn by horses; what is noteworthy, however, is that these particular horses are characterized as being "*polyphrastoi*," meaning literally "much-intelligent" ("*die vielverständigen Rosse*" in the German of Hermann Diels).[21] Hinze perceives here a parallel to the *Kundalinī-Śakti*, which is also "much-intelligent" or "*vielverständig*." He qualifies this particular concordance, however, as falling short of "identity," since "the full equality of the metaphor is lacking here"; yet he goes on to propose a list of concordances "which I do not hesitate to view as exact parallels." He notes, first of all, that the awakening of *Kundalinī* is associated with a threefold experience, involving heat, sound and a rotating movement, all of which are mentioned in the text ("And the axle in the hubs let out the sound of a pipe blazing from the pressure..."). Next he points out that the "maidens" who "lead the way" correspond to the *Śaktis* in the respective *cakras*. Where the poem speaks of the "much-famous road of the divinity that carries the man who knows through the vast and dark unknown," Hinze perceives "the royal way" of the *suṣumnā nāḍī*, "which also lies outside the normal sphere of humans, but nevertheless enjoys great renown." To the "paths of Day and Night" correspond the *nāḍis Piṅgalā* and *Iḍā*, "which are not only designated by the same names but also play the same role." And where Parmenides speaks of an "ethereal" gate at which Day

---

21. *Die Fragmente der Vorsokratiker*, vol. 1 (Zürich: Weidmann, 1968), p. 228.

and Night are joined, Hinze perceives the fifth *cakra*, named *viśuddha*, situated in the throat and associated with the element *ākāśā*, that is to say, the "ether." It is said in Tantric tradition that this *cakra* does in fact constitute a "juncture of Day and Night"—that is, of *Pingalā* and *Idā*—and thus constitutes a center at which the polarizations of the nether world are in fact transcended. "The Yogi is able here to view together the past, the present, and the future," a power known as *trikāla jñāna siddhi*. Situated in the throat, this *cakra* constitutes the Gate, known as "the Door of the Great Liberation," that leads directly into the spiritual or celestial world: "from the unreal (*asat*) to the real (*sat*), from darkness to light, from death to immortality" as a famous Vedic prayer declares. And we may ask ourselves: is this not also perhaps "the narrow gate" of the Gospels, "the eye of the needle" through which "camels" cannot pass? Entering through this Gate the Yogi reaches the *ājñā-cakra*, often depicted as a "third eye" at the center of the forehead, by which he perceives the highest tier of the *tribhuvana* or "triple world." As Hinze explains:

> Here also is the place where the Goddess, who teaches Parmenides the doctrine of Being and Non-Being, receives him. *The entire doctrinal poem is characteristic of this sphere,* is in fact an exact representation of the truth as comprehended from this particular level. The experience which Parmenides had attained was not the supreme realization of Being *(nirvikalpa samādhi)* in the thousand-petalled lotus but the "restricted" realization of Being *(savikalpa samādhi)* in the region of the *ājñā-cakra.* (109)

In point of fact, as Hinze goes on to observe, in the *sahasrāra* "there is no longer scope for speech," since sound (*śabda*) originates below that level: in the *ājñā-cakra*, namely.

Such, in brief, is Hinze's "yogic" interpretation of the Journey recounted by Parmenides: of its means, its destination, and its purpose. One must not suppose, however, that because this Journey terminates *below* the level of *sahasrāra*, the resultant doctrine is *ipso facto* provisional, imperfect or incomplete; the point, rather, is that *doctrine as such* is, in a way, subordinate to the truth it expresses, which is something else entirely. What stands at issue is precisely the distinction between *gnosis*, properly so called, and *doctrinal gnosis*, which on the one hand constitutes an expression or manifestation of *gnosis* in linguistic mode, and on the other a sign which can serve as a "means" to its attainment.[22] Now, doctrinal gnosis itself admits both modes and degrees; there is a categorical distinction to be made, for example, between *oral* and *written* transmission. Granting, then, that there are different kinds and levels of doctrinal gnosis, where is the doctrine of Parmenides to be placed? Suffice it to say that, *qua* doctrine "orally conveyed by the Goddess herself," it stands higher than all human philosophy, and that, as "an exact representation of the truth" pertaining to the highest sphere at which there is yet "scope for speech," it evidently stands supreme.

It is to be noted that much has been learned in recent times regarding the ancient Phocaeans, the forebears of Parmenides, and that the newly-discovered facts tend to confirm Hinze's conclusions. Already the name is significant, since it derives from the word *"phoca,"* which means "seal": an *amphibious* animal, namely. It appears that the Phocaeans did practice certain disciplines that may be characterized as "yogic," and that they were known especially for their powers of healing and their "journeys into other worlds." It is said that they were given to the practice of *hesychia*, a discipline of silence or stillness, and could enter into states of suspended animation; one wonders whether the "hesychasm" observed to this day at Mount Athos might not be ultimately of Phocaean origin in its technical aspects. It is moreover of great significance that an inscription unearthed at Velia, the birthplace of Parmenides, refers to him as a "son of Apollo," implying that Parmenides was indeed an initiate, someone who, in the language of the time, was called a *iatromantis*. What I find remarkable is not that such was the case, but that this fact—and all that it entails!—could have been forgotten

---

22. On this subject I refer to Chapter 1 of my monograph *Christian Gnosis: From Saint Paul to Meister Eckhart* (Tacoma, WA: Angelico Press/Sophia Perennis, 2012).

in the course of barely two centuries, and subsequently left out of account. But as one may surmise, such a Lethean "forgetting" did come to pass, and seems indeed to take place invariably at the termination of an era, the birth of a "new age." Suffice it to say that what historians admiringly term "the Greek miracle" came about in the wake of the Presocratics, when the ways and the wisdom of the old masters sank into oblivion, and a brand new *Zeitgeist* began to assert itself.

This does not mean, however, that the teaching of Parmenides was simply abandoned or ignored; on the contrary, the doctrine stood at the center of the ensuing philosophic ferment for a century or two, and it may not be too much to say that the schools that emerged, from Plato to the Sophists, arose in reaction to the enigmatic words of the Goddess. One way or another, the Parmenidean doctrine had to be dismembered or "slain"; and in point of fact, Plato himself refers to this act as a "parricide."[23] But these are matters far beyond the scope of our present focus[24]; what I wish to convey is simply that "the true Parmenides" has been hidden from view for more than two thousand years.

In conclusion, I would point out that the Phocaeans, in addition to their "mystical" pursuits, took great interest in both astronomy and geography; and one may presume that Parmenides must have been privy not only to their "yogic" practices, but to their scientific findings as well. It should be recalled that what is perhaps the first major scientific discovery in history—namely, the recognition that the Earth is spherical—has in fact been attributed by ancient authors to Parmenides. This particular insight pertains thus to Part Two of his didactic poem, the portion of his *magnum opus* which has generally been neglected or somehow "explained away." Scholars inform us that the oldest fully extant text which speaks of the Earth as spherical is indeed the *Phaedo*; but Plato himself makes it clear that this teaching has been passed on to him from older

23. *Sophist*, 241b.

24. On this subject I refer to the writings of Jean Borella. See especially *Penser l'analogie* (Geneva: Ad Solem, 2000), pp. 136–61, and *La crise du symbolisme religieux* (Lausanne: L'Age D'Homme, 1990), pp. 281–304.

sources. Most significantly, however, he refers to that hallowed scientific tenet as a "myth": what are we to make of this? Here is what Peter Kingsley has to say in that regard:

> We pride ourselves on being able to separate fact from fiction, science from myth, but don't see that our science itself is what it always has been: a fragile mythology of the moment.... And so we come back to the fact that in Plato's *Phaedo*—the first complete text still surviving to say the earth is a sphere—the idea of a spherical earth is presented to us fairly and squarely as a myth. For this is no coincidence. It's not the result of some bizarre accident; of some strange freak of history or nature. It's because Plato's friends had taught him well.[25]

One is beginning to see that the *aparavidyā* of Parmenides, so far from being "illusory" in the vulgar sense, does in fact comprise knowledge of a scientific kind, even by the standards of our day. As Kingsley has brilliantly observed: "To dismiss the illusion as just an illusion is itself just an illusion." Yet, though it cannot be dismissed as "just an illusion," we may presume that the factor of "illusion"— call it *avidyā*, *māyā*, "deception," or what you will—is with us nonetheless, and that this holds true for all human modes of knowing, from the simplest judgment to the kind that wins a Nobel Prize. We need to understand that the words of admonition spoken by the Goddess have lost nothing of their relevance or urgency; to quote Kingsley again:

> Well over two thousand years ago, science as we know it was offered to the West with a warning tag attached to it: Use this, but don't be tricked by it. And of course, impatient children that we are, we tore off the tag and ignored the warning.[26]

One might add that in Plato's Academy the "tag" was still in place, as evidenced by the mythical status ascribed to the notion of a "spherical Earth" in the *Phaedo*. It may have been Aristotle who "tore off the tag"; in any case, what we need above all to realize is

25. *Reality*, op. cit., p. 254.
26. Ibid., pp. 253–54.

that our science, *even at its best*, is yet an *aparavidyā*: a lesser and imperfect knowing, which for all its prowess is indeed "deceptive," just as the Goddess has said. What precisely does this mean? There is no simple answer, no answer on the "lower" plane. As darkness is invisible in itself, so the "illusory" character of the lower knowing can not be discerned on its own ground. What is needed is some participation in the higher knowing, the *paravidyā*; and this is a task for the authentic metaphysician, and ideally, for the initiate, the enlightened sage, the "son of Apollo."

7

FROM PHYSICS TO SCIENCE FICTION: RESPONSE TO STEPHEN HAWKING

CERTAINLY Stephen Hawking's latest book, *The Grand Design*,[1] is not simply another "Physics for the Millions" production, nor is Hawking himself just another scientist addressing the public at large. The appearance of this treatise is rather to be seen as the crossing of a threshold, an event comparable, in a way, to the publication of Charles Darwin's *magnum opus* a century and a half ago. There have always been physicists who make it a point, in the name of their science, to dispatch the "God-hypothesis"; what confronts us, however, in *The Grand Design* is something more. It is the spectacle of a physics, no less, explaining how the universe itself came to be: "*why there is something rather than nothing*" as Hawking declares. The answer to this supreme conundrum, we are told, can now be given on rigorous mathematical grounds by physics itself: such is the "breakthrough" the treatise proposes to expound in terms simple enough to fall within the purview of the non-specialist.

To appreciate the significance and potential impact of *The Grand Design*, we need to remind ourselves that following the demise of Albert Einstein, it is Stephen Hawking who has become, in the public eye, the première physicist: the lone figure that personifies the wizardry of mathematical physics as such. Add this fact to the brilliance of the book itself, and one begins to sense the magnitude

---

1. Published by Random House in 2010 with co-author Leonard Mlodinow.

156

of its likely impact, the effect upon millions of the claim that *a mathematical physics has trashed the sacred wisdom of mankind!*

This contention must not go unanswered. It calls for a definitive response, a rigorous refutation; and such I propose to present in the sequel with the help of Almighty God: the very God whose existence has supposedly been disproved.

The essay is divided into three parts. The first gives an overview of *The Grand Design*, chapter by chapter, setting forth its key conceptions and the overall logic of its argument. The second offers a five-fold refutation, based upon both philosophic and scientific grounds. The third, finally, seeks to place the phenomenon of Hawking's best-seller in perspective by reflecting upon the nature, motivation, and limits of the scientific enterprise as such.

I

Before embarking upon a critique of Hawking's doctrine, I propose to enunciate not just selected propositions destined to be the targets of criticism, but indeed the central ideas of *The Grand Design*. I propose, moreover, to place these tenets before the reader, not as so many isolated fragments, but so as to exhibit their function in the doctrine as a whole. Lastly, I shall endeavor not to condense this summary to the point where it loses all flavor, but to convey, apart from the bare logic of the text, a sense of its over-all brilliance, its power to enthrall: only thus can one appreciate fully what in fact stands at issue.

We begin with **Chapter 1**, entitled "The Mystery of Being," which does in fact deal with basic ontological issues. "Traditionally these are questions for philosophy," Hawking[2] writes, "but philosophy is dead. Philosophy has not kept up with modern developments in science, particularly physics. Scientists have become the bearers of the torch of discovery in our quest for knowledge." (5)[3] Following this opening salvo, Hawking begins to delineate the radical change

2. We do not wish to slight the co-author, Leonard Mlodinow, by referring to Hawking alone.

3. This shall mean that the quotation occurs on page 5 of *The Grand Design*.

in the conception of "being"—he means of course *physical* being— implied by the transition from classical to quantum physics. "According to the traditional conception of the universe, objects move on well-defined paths and have definite histories."[4] Not so in quantum theory. Availing himself of the fact that quantum mechanics can be formulated in a number of different ways which turn out to be mathematically equivalent, Hawking chooses the approach pioneered by the American physicist Richard Feynman as best suited to convey his thought. And whereas he postpones his presentation of quantum theory *à la* Feynman till Chapter 4, he forthwith makes a central point: "According to Feynman, a system has not just one history, but every possible history." (6) One sees that Hawking has started to make his case: it begins to appear that the new ontology has indeed left traditional conceptions of "being" far behind.

Noting that things are not "what they seem as perceived by the senses" (7), Hawking announces one of his foundational innovations: the concept of "model-dependent realism," which is "based upon the idea that our brains interpret the input from our sensory organs by making a model of the world." One should add that the full force of what Hawking has in mind becomes apparent in Chapter 3 with the assertion that "There is no picture- or theory-free concept of reality" (42), where also we are told that model-based realism is "the idea that a physical theory or world-picture is a model (generally of a mathematical nature) and a set of rules that connect the elements of the model to observations." (43) Getting back to Chapter 1: Following the announcement of this crucial conception, Hawking goes on to consider the history of human knowing, "from Plato to the classical theory of Newton to modern quantum theories" (7), and proceeds to pose the following question: "Will this sequence eventually reach an end point, an ultimate theory of the universe, that will include all forces and predict every observation we can make, or will we continue forever finding better theories, but never one that cannot be improved upon?" Now, it is at this juncture that Hawking breaks with his predecessor, Albert

---

4. Whenever a quotation is not followed by its page number, the preceding page number is to be understood.

Einstein: there *is no* "ultimate theory" as previously conceived which covers the entire ground, he maintains. What is called for is a radically new kind of theory, something he terms "M-theory," a notion that dovetails with the conception of "model-dependent realism"; as Hawking explains: "M-theory is not a theory in the usual sense. It is a whole family of different theories, each of which is a good description of observations only in some range of physical situations."(8) The ultimate goal of physics—a science, namely, which in principle covers the entire ground—can only be realized as an M-theory; and Hawking believes that physics today is closing in upon such a final and all-inclusive formulation.

This brings us to the most amazing claim of all: the notion that such an M-theory constitutes the culmination not only of physics, but of philosophy as well: that it is in fact the only *kind* of theory that can enlighten us regarding "the mystery of being." And what does it reveal? It informs us, first of all, that "ours is not the only universe," that indeed "a great many universes were created out of nothing." But—as if this were not enough!—there is more: the final M-theory, we are told, will in principle reveal all that *can* be known, not only regarding our universe, but indeed regarding *everything*.

The plan of the book has now come into view: it can evidently be none other than to lead the reader, step by step, through the formulation of the ultimate M-theory, as far as Hawking can take us at this time.

Chapter 2 deals with "The Rule of Law." It begins with a quotation from Viking mythology concerning wolves that pursue the sun and the moon, the point being that when they catch either one, there is supposedly an eclipse. "Ignorance of nature's ways," Hawking concludes (following several more such examples), "led people in ancient times to invent gods to lord it over every aspect of human life."(17) After informing us that "Our species, *Homo sapiens*, originated in sub-Saharan Africa around 200,000 BC," Hawking proceeds to trace the first rudimentary beginnings of scientific enlightenment: the recognition, however dim and distorted, of "the Rule of Law." The first phase of this human evolution proceeds from Thales of Miletus and Pythagoras to Anaximander, Empedocles,

Aristarchus, and Ptolemy; next come the Middle Ages, the Renaissance, and the beginning of the modern age, in which *science*, properly so called, comes at last to birth, thanks to the labors of Kepler, Galileo, and Descartes. There is however no need to summarize this account, which in fact does not differ substantially from the customary expositions. Suffice it to note that "The modern concept of laws of nature emerged in the seventeenth century. Kepler seems to have been the first scientist to understand the term in the sense of modern science." (25) As regards Galileo, not only did he "uncover a great many laws," but he "advocated the important principle that observation is the basis of science, and that the purpose of science is to research the quantitative relationships that exist between physical phenomena." (26) Descartes comes next; and here the account focuses upon the Cartesian conception of "law" and the notion of "trajectories" uniquely determined by their initial conditions. The stage is now set for Newton, whose epochal achievements Hawking barely touches upon at this juncture; they are to be considered later, in their relation to post-Newtonian physics.

True to its title, the chapter is indeed focused upon "the Rule of Law." There are, in particular, three fundamental questions regarding that Rule the author wishes to consider: first, "What is the origin of the laws?"; secondly, "Are there any exceptions to the laws, i.e., miracles?"; and thirdly, "Is there only one set of possible laws?" As the reader may have surmised by now, these are among the issues Hawking proposes to resolve on the basis of M-theory. For the moment, however, his concern is with the second: the question of physical determinism. And on this issue he cites Laplace as the great inaugurator: "The scientific determinism that Laplace formulated is the modern scientist's answer to question two. It is, in fact, the basis of all modern science, and a principle that is important throughout this book." (30). To be precise, the principle affirms that "Given the state of the universe at one time, a complete set of laws fully determines both the future and the past." It is to be noted that there appears to be a conflict between "scientific determinism" as thus conceived and what is commonly referred to as quantum-mechanical "indeterminism," a question Hawking will address in Chapter 4.

But let us continue. No sooner has he formulated the notion of universal determinism than he observes: "Since people live in the universe and interact with other objects in it, scientific determinism must hold for people as well." And to be sure, this means that in reality *there is no such thing as "free will."* As Hawking goes on to explain: "Though we feel that we can choose what we do, understanding of the molecular basis of biology shows that biological processes are governed by the laws of physics and chemistry and therefore are as determined as the orbits of the planets." (32) Indeed, "Recent experiments in neuroscience support the view that it is our physical brain, following the known laws of science, that determines our actions, and not some agency that exists outside those laws." And of course this implies that there can be no free will: "It is hard to imagine how free will can operate if our behavior is determined by physical law, so it seems that we are no more than biological machines and that free will is just an illusion."

Certainly Hawking admits the impossibility of actually calculating human behavior; but this means, not that the human organism fails to reduce to a physical system, but that it is far too complex a system to be tractable. "Because it is so impractical to use the underlying physical laws to predict human behavior," Hawking goes on, "we adopt what is called an effective theory. In physics an effective theory is a framework created to model certain observed phenomena without describing in detail the underlying processes." So too, in the case of persons, we can speak of "free will" on the level of an effective theory: "The study of our will, and of the behavior that arises from it, is the science of psychology." (33)

Thus we arrive, finally, at the conclusion of the chapter, the fact that "This book is rooted in the concept of scientific determinism." (34)

In **Chapter 3** ("What is Reality?") Hawking explores the scientific implications of model-dependent realism. He begins by contrasting Ptolemaic geocentrism with Copernican heliocentrism, and concludes that "Although it is not uncommon for people to say that Copernicus proved Ptolemy wrong, that is not true." (41) The point is that "one can use either picture as a model of the universe"; it is only that the "the equations of motion are much simpler in the frame of reference in which the sun is at rest." (42) And this brings us to the central premise: "*There is no picture- or theory-independent concept of reality.*" It is to be noted that this seemingly innocuous notion has profound implications; for it means that a scientific theory is not a description of an independently-existing reality (as scientists and laymen alike had thought), but a "model" that *defines* reality. According to model-dependent realism, the concept of a *model-independent* reality proves to be vacuous. What happens now if different models agree with the corresponding observations? "If there are two models that both agree with observation," Hawking maintains, "then one cannot say that one is more real than another." (46) In effect, one can identify the two model-dependent realities, even as we habitually identify two views of a solid object corresponding to different points of observation.

To the question why the classical (or "model-independent") realism was abandoned Hawking gives an answer based upon quantum theory: "Though [classical] realism may be a tempting viewpoint, as we'll see later, what we know about modern physics makes it a difficult one to defend. For example, according to the principles of quantum theory, which is an accurate description of nature, a particle has neither a definite position nor a definite velocity unless and until those quantities are measured by an observer." (44) Yet Hawking does not rest content with a new philosophy of physics, but affirms that the idea of model-dependent realism applies also, as we have seen (in reference to Chapter 1), to pre-scientific ways of knowing, inclusive of sense perception: "Model-dependent realism," he reiterates, "applies not only to scientific models but also to the conscious and subconscious mental models we all create in order to interpret and understand the everyday world." (46) And he goes on to emphasize: "There is no way to remove the observer— us—from our perception of the world, which is created through our sensory processing, and through the way we think and reason." He then speaks of perception, of the signals sent along the optic nerve to the brain, and the processing that takes place within that organ, for example, the construction of a third dimension not given in the retinal image: "The brain, in other words, builds a mental picture or model. . . . This shows that what one means when one says 'I see a chair' is merely that one has used light scattered by the chair to build a mental image or model of the chair." (47)

Next Hawking addresses the likely question whether "things"—for instance, tables—"exist" when they are not perceived. And his solution is simple: "The model in which the table stays put is much simpler, and agrees with observation. That is all one can ask." The same logic applies to fundamental particles, which cannot be perceived, but yet can be "observed": electrons, for example, "exist" even before they affect an instrument of detection (such as a television screen). The case of quarks (believed to be the components out of which protons, neutrons and pi-mesons are formed) is a bit more complicated, because "individual" quarks cannot be observed; but logically the case stands the same: the model in which quarks exist "is much simpler, and agrees with observation. This is all one can ask."

Although some models have greater explanatory power than others, Hawking insists that they cannot be said to be more "real" (51), presumably because it makes no sense to quantify or otherwise "rank" model-dependent realities. He thus compares the biblical account of cosmogenesis with big bang cosmogony, which "explains the fossil and radioactive records and the fact that we receive light from galaxies millions of light-years from us," and is consequently "more useful than the first." Yet, even so, "neither model can be said to be more real than the other."

At this point one senses the need for criteria which enable one to rank theories, to determine how "good" a model is; and we will mention, in passing, that Hawking gives four: i.e., whether a theory "is elegant," whether it "contains few arbitrary or adjustable elements," whether it "agrees with and explains all existing observations," and whether it "makes detailed predictions about future observations that can disprove or falsify the model if they are not borne out."

This brings us finally to the crucial notion of "dualities" which Hawking introduces near the end of the chapter. He cites the example of "wave-particle duality": the fact that light, for instance, can be described or "modeled" in both wave and particle terms. "Dualities like this—situations in which two very different theories accurately describe the same phenomenon—are consistent with model-dependent realism." (58) The point proves to be decisive for the following reason:

> There seems to be no single mathematical model or theory that can describe every aspect of the universe. Instead, as mentioned in the opening chapter, there seems to be the network of theories called M-theory.... Wherever their ranges overlap, the various theories in the network agree, so they can be said to be parts of the same theory.... Though this situation does not fulfill the traditional physicist's dream of a single unified theory, it is acceptable within the framework of model-dependent realism.

Chapter 4 ("Alternative Histories") begins with a description of the famous "double-slit" experiment, which according to Richard Feynman "contains all the mystery of quantum mechanics." The idea goes back to an experiment performed in the nineteenth century by Thomas Young, in which light was passed through a screen with two slits to a surface behind the screen. This gave rise, not simply to a single bright line behind each slit, but to a pattern of bright and dark regions, of multiple "lines." There is however no mystery here: given that light consists of waves (as most scientists had surmised from the start), these "lines" are simply the pattern resulting from the fact that when two waves are superposed, the resultant amplitude attains a maximum whenever "crest meets crest," and a minimum when "crest meets trough." What has astounded physicists, on the other hand, is that the same happens when the experiment is conducted with particles instead of waves.[5] What is critical is the *size* of the particles: the effect ceases to be measurable with particles large enough to be perceptible.[6] What is perhaps most baffling of all is that the effect persists even if the particles in question are passed

---

5. The first experiment of this kind was carried out in 1927 by two physicists at Bell Labs using electrons.

6. As Hawking relates, the largest particles used thus far (in an experiment conducted in Austria in 1999) are certain molecules, called "buckyballs," made up of 60 carbon atoms.

through the slit "one at a time": one finds that so long as both slits are open, the interference pattern remains. In some mysterious way an electron, say, passing through slit A, "knows" whether slit B is open or closed. This alone makes it clear that, on an atomic or sub-atomic scale, the conceptions and laws of classical physics break down: and that is where quantum theory comes into play, a physics which does in certain ways treat particles as waves.

Following this fundamental recognition, Hawking goes on to expound the basic ideas that differentiate quantum physics from Newtonian mechanics, beginning with the Heisenberg "uncertainty principle," which affirms that certain pairs of variables, such as the position and velocity of a particle, cannot be measured with perfect accuracy: the more precisely we know one of these variables, the greater will be the "uncertainty" pertaining to the other. In fact, according to quantum theory, an electron, say, *does not have* simultaneously a precise position and velocity: observables remain somehow diffuse or "ghostlike" unless an act of measurement limits their dispersion.

One sees that Heisenberg uncertainty entails the breakdown of the classical determinism; as Hawking informs us, "the outcome of physical processes cannot be predicted with certainty because they are not *determined* with certainty." (72) Nature "does not dictate the outcome of any process or experiment, even in the simplest of situations. Rather, it allows a number of different eventualities to be realized, each with a certain likelihood of being realized."[7] One is struck by the fact that this admission seems to contradict the Laplacian principle of scientific determinism, enunciated in Chapter 2 as "the basis of modern science" (30), which asserts that "given the state of the universe at one time, a complete set of laws *fully determines*[8] both the future and the past"! Not so, Hawking maintains: "Quantum theory might seem to undermine the idea that nature is

7. However, when it comes to the macroscopic processes to which classical physics applies, the resultant probability distribution for the outcome of a measurement is so sharply concentrated around its mean as to determine a unique value within the accuracy of measurement. In other words, quantum theory reduces in effect to classical physics in the macrocosmic domain.

8. My emphasis.

governed by laws, but that is not the case. Instead it leads us to accept a new form of determinism: Given the state of a system at some time, the laws of nature determine the probabilities of various futures and pasts rather than determining the future and past with certainty." For most scientists, admittedly, this was an unwelcome admission, and only in the face of incontrovertible evidence did they eventually accede to it: Laplace notwithstanding, there *is* finally no "complete set of laws" that "*fully determines* both the future and the past."

Despite the probabilistic nature of quantum mechanical predictions, however, its claims can be rigorously tested, which is to say that probability distributions can be observed by statistical means. Quantum theory is still physics: a rigorous science which gives rise to quantitative predictions that can be verified or falsified by experiment; and as Hawking points out: "It has never failed a test, and it has been tested more than any other theory of science." (74)

He goes on to point out that the probabilities of quantum theory are of a kind unknown in everyday life. The toss of a coin, for example, gives rise to a probability distribution, not because it is intrinsically indeterminate, but simply because we cannot control the parameters descriptive of the toss with sufficient accuracy to determine the resultant trajectory. "Probabilities in quantum theories," however, "are different. They reflect a fundamental randomness in nature." What stands at issue has puzzled the greatest physicists— and especially the greatest, one might add—from Albert Einstein to Richard Feynman, who brooded over this "fundamental randomness" for years, and was lead finally to observe: "I think I can safely say that nobody understands quantum mechanics."

Hawking turns now to a formulation of quantum mechanics introduced by Feynman in the 40's, which "has proved more useful than the original one." (76) It is based upon an exceedingly bold idea, the kind only a scientific genius of first rank can successfully bring into play. Consider the double slit experiment, carried out with particles of some kind. One knows from quantum theory that a particle has no definite position between the moment it embarks upon its path and the moment it is detected at the second screen. But instead of interpreting this to mean that particles "take *no* path as they travel between the source and the screen," Feynman realized that it could mean instead that "particles take *every* possible path connecting those points." Herein, he felt, lies the secret of quantum theory: "This, Feynman asserted, is what makes quantum physics different from Newtonian physics." (75) And since "Feynman's view of quantum reality is crucial in understanding the theories we will soon present," Hawking makes it a point to give us "a feeling for how it works." (77)

Consider the double-slit experiment. To determine the probability amplitude for a particle at a point A on the second screen, we need to add the contribution to that amplitude for every path from the source O to A. Now, what matters is the phase contributed by any given trajectory (for example, whether the corresponding wave has a crest or a trough at A), and what renders this calculable is the fact that for all but special paths, the contributions from nearby paths cancel.[9] These ideas, however, carry over from the case of the double-slit experiment to the general case of a particle moving from one point to another: "Feynman's mathematical prescription . . . showed that when you add together the waves from all the paths you get the 'probability amplitude' that the particle, starting at A, will reach B." The same holds true, moreover, for an arbitrary physical system composed of any number of particles: "Feynman showed that, for a general system, the probability of any observation is constructed from all the possible histories that could have led to that observation. Because of that his method is called the 'sum over histories' or 'alternative histories' formulation of quantum physics." (82)

Having thus introduced the reader to Feynman's version of quantum theory based upon the notion of "alternative histories," Hawking touches upon another "strange" feature of the new physics, the fact that "the (unobserved) past, like the future, is indefinite and exists only as a spectrum of possibilities. The universe, according to quantum physics, has no single past, no history." And this implies (what is perhaps the weirdest fact of all!) "that observations you

---

9. To speak of "phase" and "cancellation" is of course to speak in terms of a wave representation. We must recall that in quantum theory particles are also treated as waves.

make on a system in the present affect its past." Such so-called "delayed choice" experiments can be carried out, for example, in the case of the double-slit scenario. But Hawking is mainly concerned to pursue the notion of "delayed choice" to its ultimate conclusion: "We will see that, like a particle, the universe doesn't have just a single history, but every possible history, each with its own probability; and our observations of its current state affect its past and determine the different histories of the universe, just as the observations of the particles in the double-slit experiment affect the particle's past." (83)

Chapter 5 ("The Theory of Everything") commences with an overview of post-Newtonian classical physics, beginning with the discovery of the electromagnetic field culminating in the field equations of James Clerk Maxwell. All manner of electromagnetic waves, from X-rays to visible light to radio waves, could now be described with unprecedented accuracy. A fundamental difficulty, however, presented itself: it was assumed that the electromagnetic field presupposed a medium permeating all space, the so-called ether, a tenet which has scientific implications: "If the ether existed, there would be an absolute standard of rest ... and hence an absolute way of defining motion as well. The ether would provide a preferred frame of reference throughout the entire universe, against which any object's speed could be measured." (93) In conjunction with the Galilean hypothesis of a stationary sun, around which the Earth revolves with an orbital velocity v (relative to the ether), one is led to ask whether it may be possible to measure v. In 1887, moreover, Albert Michelson and Edward Morley did in fact conduct such an experiment, based upon the following idea: if c designates the velocity of light (relative to the ether), then its velocity relative to the Earth should be $c-v$ for a light beam moving in the same direction as the Earth, and $c+v$ when it moves in the opposite direction. However, the experiment disclosed—to the consternation of the scientific community!—that the two relative velocities are in fact equal.[10]

---

10. Hawking stops short of pointing out what this means: it implies (on the basis of Newtonian physics) that $v = 0$, which is to say that, contrary to the Galilean tenet, the Earth does *not* move. We will return to this point in Part III.

At this critical juncture Hawking proceeds to delineate the basic conceptions of Einsteinian relativity, beginning with the special theory (published in 1905), which resolves the aforesaid impasse by stipulating that the velocity of light is one and the same in every so-called inertial frame of reference. This leads mathematically to the notion of a 4-dimensional space-time continuum, and to a corresponding modification of the Newtonian equations. The special theory was then extended (in 1917) to arbitrary frames of reference in the so-called general theory of relativity, which is based upon the revolutionary notion that gravitational fields can be explained *geometrically* as resulting from a "curvature," not of the now discarded 3-dimensional space, but of 4-dimensional space-time, precisely. In brief but intuitively comprehensible terms Hawking pilots us through this development, an exposition which concludes with the claim that Einsteinian relativity (inclusive of the general theory) has meanwhile been verified by an array of experiments, ranging from measurements by atomic clocks mounted on airplanes circling the Earth, to data derived from GPS satellites said to detect "gravitational" effects. "Modern technology," Hawking informs us, "is sensitive enough to allow us to perform many sensitive tests of general relativity, and it has passed every one." (102)

Hawking's vision of physics, however, differs radically from that of Einstein; like the Maxwellian theory which it has replaced, Einsteinian physics itself is not the last word: "Though they both revolutionized physics, Maxwell's theory of electromagnetism and Einstein's theory of gravity—general relativity—are both, like Newton's own physics, classical theories. That is, they are models in which the universe has a single history. As we saw in the last chapter, at the atomic and subatomic levels these models do not agree with observation." (103) What is needed, Hawking contends, is a quantum theory which embraces not only Newtonian mechanics, but the electromagnetic theory of Maxwell and Einstein's gravitational theory as well. To be precise, there are four basic forces of nature: gravity, electromagnetism, and the so-called weak and strong nuclear force. Now, quantum mechanics as originally conceived (around 1925) was essentially a theory of matter: of mass particles, that is, such as protons, neutrons and electrons. What is now called for, to complete the picture, is a quantum theory in which not only matter but also force fields are "quantized," that is to say, treated from a quantum theoretic point of view. This is where the so-called quantum field theories enter the picture; as Hawking explains, "in quantum field theories the force fields are pictured as being made of various elementary particles called bosons, which are force-carrying particles that fly back and forth between matter particles, transmitting the forces. The matter particles are called fermions." (104)

The first field to be successfully quantized was the electromagnetic, resulting in quantum electrodynamics or QED, a theory evolved in the 40's with Feynman in the lead. The first boson, thus, to be discovered, was the photon: "According to QED all the interactions between charged particles—particles that feel the electromagnetic force—are described in terms of an exchange of photons." (105) And one might add that QED ranks among the most spectacularly accurate physical theories yet devised.

Before proceeding to the next feat of field quantization, Hawking touches upon two brilliant conceptions, both introduced by Feynman, that render such quantization possible. The first pertains to the so-called "Feynman diagrams" which enable one to calculate the aforesaid "integrals over histories" entering into the formalism of quantum field theories, diagrams which Hawking regards as "one of the most important tools of modern physics." A second hurdle that needed to be overcome was the daunting fact that "When you add the contributions from the infinite numbers of different histories, you get an infinite result." (107) And this is where another of Feynman's master-strokes comes into play: to deal with this fundamental difficulty, he invented a mathematical procedure termed "renormalization." The process involves "subtracting quantities that are defined to be infinite and negative in such a way that, with careful mathematical accounting, the sum of the negative infinite values and the positive infinite values that arise in the theory cancel out, leaving a small remainder, the finite observed values of mass and charge."

As Hawking points out, it was this breakthrough, achieved in QED, that encouraged physicist to attempt the quantization of other fields. It eventually became apparent, however, that to this end these fields had to be somehow unified: one began to surmise that "the division of natural forces into four classes is probably artificial and a consequence of our lack of understanding." (109) And thus began the search for "a theory of everything that will unify the four classes into a single law that is compatible with quantum theory." A first breakthrough in that regard was achieved in 1967, when Abdus Salam and Steven Weinberg "each independently proposed a theory in which electromagnetism was unified with the weak force, and found that the unification cured the plague of infinities. The unified force is called the electroweak force. Its theory could be renormalized, and it predicted three new particles, W^+, W^-, and Z^o." The search for these particles was now on at major nuclear research facilities, and by 1983 all three were found to exist.

Next came the strong nuclear force. "The strong force can be renormalized on its own in a theory called QCD, or quantum chromodynamics. According to QCD, the proton, the neutron, and many other elementary particles of matter are made of quarks, which have a remarkable property that physicists have come to call color." This curious nomenclature (which obviously must not be taken literally) serves to label the three kinds of quarks predicted by the theory: they are characterized as "red, green, and blue." The next step towards unification consisted in the formulation of so-called grand unified theories or GUT's, which attempted to unify the strong and electroweak forces; but these attempts have proved unsuccessful: in consequence of adverse observational evidence "most physicists adopted an ad hoc theory called the standard model, which comprises the unified theory of the electroweak forces and QCD as a theory of the strong force… The standard model is very successful and agrees with all current observational evidence, but it is ultimately unsatisfactory because, apart from not unifying the electroweak and strong forces, it does not include gravity." (112)

It is here, in its encounter with gravity, that quantum field theory runs into its greatest obstacle. In consequence of Heisenberg uncertainty the gravitational field cannot maintain its state of minimum energy, called the vacuum, without "what are called quantum jitters, or vacuum fluctuations—particles and fields quivering in and out of existence." (113) These phantom particles, which occur in pairs, are called "virtual," and despite the fact that they cannot be directly observed, their effects upon electron orbits, though exceedingly small, "can be measured, and agree with theoretical predictions to a remarkable degree of accuracy." There is however a major problem, which is that "the virtual particles have energy, and because there are an infinite number of virtual pairs, they would have an infinite amount of energy. According to general relativity, this means that they would curve the universe to an infinitely small size, which obviously does not happen!"

It is this impasse that has prompted another major conceptual leap, perhaps the most gigantic of all. The new theory, proposed in 1976, is termed supergravity, a designation in which the prefix refers to "a kind of symmetry the theory possesses, called supersymmetry," which implies that "force and matter particles, and hence force and matter, are really just two facets of the same thing. Practically speaking, that means that each matter particle, such as a quark, ought to have a partner particle that is a force particle, and each force particle, such as a photon, ought to have a partner particle that is a matter particle." (114) The problem is that as yet "no such partner particles have been observed" (115), due perhaps to the fact that these particles are supposed to be about a thousand times heavier than the proton; "but there is hope that such particles will eventually be created in the large Hadron Collider in Geneva."

It happens, moreover, that the idea of supersymmetry antecedes the theory of supergravity, having originated in the so-called string or "superstring" theories. What is most striking in this entire conglomerate of theories is the fact that supersymmetry requires at least ten space-time dimensions "in place of the usual four": how, then, does one get from ten or more to four? "In string theory the extra dimensions are curled up into what is called the internal space, as opposed to the three-dimensional we experience in ordinary life. As we'll see, these internal states are not just hidden dimensions swept under the rug—they have important physical significance." (116)

What is likewise of major importance is the fact that "string theorists are now convinced that string theories and supergravity are just different approximations to a more fundamental theory, each valid in different situations"; and as might now be expected, "that more fundamental theory is called M-theory. . . ." (117) It is here, precisely, that Hawking proposes his radical innovation: "It could be," he tells us, "that the physicist's traditional expectation of a single theory of nature is untenable, and there exists no single formulation." His point is that a family of theories or "models" which "agree in their predictions whenever they overlap" could do just as well. Hawking admits that we do not yet know for certain whether M-theory might not in the end turn out to be "classical," although he evidently regards this as unlikely. In any case, we do know certain facts: "First, M-theory has eleven dimensions, not ten." In addition, one knows that "M-theory can contain not only strings but also point particles, two-dimensional membranes, three-dimensional blobs, and other objects that are more difficult to picture and occupy even more dimensions, up to nine." (118) Most importantly, it is known that the constitution of the internal space determines both "the values of the physical constants, such as the charge of the electron, and the nature of the interactions between elementary particles. In other words, it determines the apparent laws of nature," that is to say, the laws we discover by empirical means. "But the more fundamental laws are those of M-theory." In fact: "The laws of M-theory therefore allow for *different universes* with different apparent laws, depending on how the internal space is curled. M-theory has solutions that allow for many different internal spaces, perhaps as many as 10^{500}, which means it allows for 10^{500} different universes, each with its own laws."

This brings us to **Chapter 6**, entitled "Choosing Our Universe." It begins with an account of big bang theory, tracing the major steps of its development, from the early contributions of Einstein, Hubble and Friedmann, through its various stages up to "inflation" theory, which claims to reduce the origin of our universe to a "quantum event." A map of the sky (on page 138), based upon data collected over seven years and released in 2010—in which a myriad dots of various colors purport to represent temperature differences of less than a thousandth degree Centigrade some 13.7 billion years ago!—concludes the presentation. "So look carefully at the map of the microwave sky," Hawking observes. "It is the blueprint for all the structure in the universe. We are the product of quantum fluctuations in the very early universe. If one were religious, one could say that God really does play dice." (139)

And now begins the most original part of Hawking's theory. "The usual assumption in cosmology is that the universe has a single definite history. One can use the laws of physics to calculate how this history develops in time. We call this the 'bottom-up' approach to cosmology." Hawking disapproves of this approach on the grounds that it presupposes a unique starting point of cosmic evolution: "Instead, one should trace the histories from the top down, backward from the present time." What Hawking objects to is the notion that the universe *has* "a unique observer-independent history." He argues instead that it is *we* who determine or "choose" our history by the fact that we inhabit *this* universe. There *may be* other histories leading to universes other than ours; and in fact, M-theory tells us that this is indeed the case.

"An important implication of the top-down approach is that the apparent laws of nature depend on the history of the universe." (140) Consider the dimension of the universe: why is space in our universe three-dimensional, when according to M-theory it could have up to ten dimensions? "The Feynman sum allows for all these [possibilities], for every possible history of the universe, but the observation that our universe has three large space dimensions selects out the subclass of histories that have the property that is being observed." (141) Hawking makes it a point, moreover, to emphasize that this is not mere speculation, not indeed science fiction, as one might suppose, but physics of the most solid kind. In fact, "The theory we describe in this chapter is testable." What Hawking has in mind, especially, is the magnitude and distribution of irregularities in the microwave background, which are among the features of our universe that have now come within range of observation, and have in fact "been found to agree exactly with the demands of inflation theory."[11] (143) More precise measurements, however, "are needed to fully differentiate top-down theory from

11. The term "inflation theory" refers to a quantum-mechanical model said to describe the early universe some 10^{-35} seconds after the initial singularity or "big bang."

others, and to either support or refute it." Be that as it may, Hawking leaves us with the belief that *our* universe stems from a "quantum event" which took place some 13.7 billion years ago.

This brings us to **Chapter 7**, "The Apparent Miracle," which addresses the question why the universe proves to be habitable, to carry a "human-friendly design." Traditionally, of course, mankind has believed that this "human-friendly design" derives from the fact that the world was created by a benevolent God; but Hawking takes issue with that belief. "The many improbable occurrences that conspired to enable our existence," he tells us, "would indeed be puzzling if ours were the only solar system in the universe." (153) But given the fact that there are billions of stars in our universe, many of which have solar systems, the hypothesis of "design" begins to become questionable. "Obviously, when the beings on a planet that supports life examine the world around them, they are bound to find that their environment satisfies the conditions they require to exist." And therein, precisely, lies the key to the apparent mystery: "It is possible to turn the last statement into a scientific principle: Our very existence imposes rules determining from where and at what time it is possible for us to observe the universe."

What Hawking has enunciated at this point is the so-called anthropic principle, or "weak anthropic principle," to be exact, concerning which much has been written in recent decades. He points out that the principle proves to be scientific in the sense that it leads to predictions which are testable, and in fact prove to be true; for example, it implies, as Robert Dicke was the first to show, that "the universe must be about 10 billion years old," which agrees quite well with the more accurate 13.7 billion figure of big bang theory.

The mystery, however, has not yet been resolved; for it happens that our existence requires not only the right kind of sun and a man-friendly planetary system, but also, on a more fundamental level, the right physical laws and constants of nature, a fact which a mere "principle of selection" cannot seem to explain. It is one thing, obviously, to "select" a friendly planetary system, and quite another to select a value of the fine structure constant that allows organic chemistry to happen. Now, it is at this juncture, precisely, that

Hawking brings something new to the table: the notion, namely, that ours is only one of some 10^{500} universes, each with its own laws; for indeed, on this basis our existence serves to "select" the physical laws of nature just as it selects our position within the space-time of the universe in which we find ourselves. Thus, by way of M-theory, Hawking has apparently justified what had come to be known as the *strong* anthropic principle, which affirms that "the fact that we exist imposes constraints not just on our *environment* but on the possible *form and content of the laws of nature* themselves." (155)

We need not follow Hawking as he relates "the tale of how the primordial universe of hydrogen, helium, and a bit of lithium evolved to a universe harboring at least one world of intelligent life": it is essentially the familiar account which begins with big bang astrophysics and culminates in the Darwinist scenario of evolution. What is presently of interest is that the laws and universal constants of nature need to be "fine-tuned" to permit the astrophysical and Darwinist phases of this process to take place. Consider, for example, the fact that life on earth is carbon-based, and that the formation of a carbon nucleus results from the so-called triple alpha process, involving a three-particle collision, the likelihood of which would be vanishingly small unless the strong nuclear force were within 0.5 percent of its observed value, the electric force within 4 percent, and so forth. Or to give another example: the existence of life on a planet requires extreme stability of its orbit; however, "it is only in three dimensions that stable elliptical orbits are possible." (160) Here, then, is the reason, Hawking argues in effect, why in *our* universe, space has *three* dimensions, instead of five or nine.

The logic of Hawking's argument is crystal clear: once the single universe of bygone days has been replaced by a veritable "multiverse," the fine-tuning of natural laws and constants can be explained by the weak anthropic principle, which is to say that the "apparent miracle" has disappeared: "the multiverse concept can explain the fine-tuning of physical law without the need for a benevolent creator who made the universe for our benefit." (165)

Even this "debunking of the God-hypothesis," however, is not yet the last word: in the final chapter (entitled "The Grand Design") Hawking proposes to answer the "*why?* questions" posed at the start of the book: *Why is there something rather than nothing? Why do we exist? Why this particular set of laws and not some other?"* (171) The substance of the chapter, to which we will confine our summary, is given in the concluding paragraphs; and as might be expected, the answer to the three *"why?* questions" derives from M-theory and the corresponding version of the anthropic principle. "Spontaneous creation [that is to say, creation conceived *à la* M-theory as a quantum event] is the reason there is something rather than nothing, why the universe exists, why we exist." (180) This is Hawking's answer to the first two questions; and his answer to the third is M-theoretic as well. It derives from the strong "multiverse" version of the anthropic principle, which explains why we encounter *"this particular set of laws and not some other."* The answer to the ultimate questions may thus be supplied by the physics now in progréss: "If the theory is confirmed by observation, it will be the successful conclusion of a search going back more than 3,000 years. We will have found the grand design." (181)

II

The first point to be made by way of response refers to the nature of science as distinguished from philosophy. "Philosophy is dead," Hawking asserts, and it is now science that carries "the torch of discovery in our quest for knowledge." (5) Yet granting that a good deal of what passes for philosophy these days may indeed be "dead," the fact remains that science and philosophy as such are very different disciplines, to the point that neither can fill in for the other. As we have noted earlier in this book, there is in fact a complementarity, an opposition one can say, between philosophy, properly so called, and science when the latter is shorn of its mythology and understood for what by right it is. To indicate, however summarily, the nature of this opposition, one needs to distinguish categorically between *thought* and *language* (a distinction, incidentally, which falls into the province of philosophy alone). Briefly stated, thought is an intentional act that seeks to apprehend an object by way of a concept, which may be defined in Scholastic terms as the *form* of the act. Language, on the other hand, is something subsidiary to thought: it is its vehicle, which serves to express and communicate thought. Now, it can be said that for philosophy, thought has primacy over language, whereas for science it is the other way round. Let me recall[12] that for the philosopher, the concept is no more than a means to a trans-conceptual end, which is finally the *unmediated knowledge of the object itself*; as the Chinese might put it, concepts serve the philosopher as "a finger pointing to the moon." The scientist, on the other hand, has no interest in "the moon," nor does he know that there is such an object. For him the concept plays a very different role; for what he seeks is not a transcendent entity, but "phenomena" in the contemporary sense of that ancient term.[13] How these so-called phenomena, moreover, stand in relation to the transcendent object is a question that concerns the philosopher alone, inasmuch as the very idea of "object" in the philosophic sense is foreign to the scientist. So too the scientist's modus operandi is opposed to the philosophic: instead of "opening" the concept in quest of a transcendent object, he closes it to consolidate his grip upon phenomena. And that is where *language* enters in a foundational capacity: as Jean Borella has made clear, the *epistemic closure* of the concept by which a science is defined is effected through a *criterion of scienticity* specified on the level of linguistic or formal expression.[14]

One sees, in light of this analysis, that philosophy and science are fundamentally opposed: whereas the philosopher treats concepts as provisional aids in quest of a transcendent object, the scientist, for his part, engages in a process of epistemic closure in search of phenomena *defined or conditioned by that very process*. As I have shown elsewhere,[15] the history of physics, from its Galilean beginnings right up to the latest "multiverse" theories, exhibits the various stages of this progressive closure, which manifests itself in a

12. See Chapter 3.

13. It is to be noted that this inherently Greek term has acquired virtually the opposite of its original and etymological sense as "that which shows itself *in itself.*" It is thus indeed the philosopher, and *not* the contemporary scientist, who in truth has his eye upon the phenomenon! See Chapter 8.

14. Jean Borella, *Histoire et théorie du symbole* (L'Age d'Homme, 2004), chap. 4, art. 1.

15. Chapter 3.

concomitant recession of the corresponding objects from actual human experience, culminating in the conception of entities pertaining to universes other than our own. What concerns us at the moment, however, is not the truth or scientific validity of these theories, but the fact that the evolution of physics confirms the aforesaid opposition between science and philosophy. The upshot of these summary considerations is simply this: *to suggest that science can, even in principle, replace philosophy "in our quest for knowledge" is to exhibit a fundamental lack of comprehension regarding the nature and scope of either discipline.*

My second point of contention pertains to Hawking's conception of model-dependent realism, which is in a way reminiscent of a fundamental metaphysical principle: what in fact I have termed "anthropic realism."[16] The latter affirms that the cosmos exists, not in splendid isolation as a Kantian *Ding an sich*, but indeed "for us," that is to say, as an object of human intentionality. Man and cosmos, therefore, belong together: they form a complementarity. But is this not essentially what "model-dependent realism" affirms as well? Here too the human observer comes into play by virtue of the fact that it is he who forges the conceptions—the "models"—in terms of which reality is defined. Yet there is a difference between model-dependent and anthropic realism, which proves to be crucial: for whereas Hawking regards the human observer as a component or part of the universe,[17] anthropic realism insists that man, the authentic anthropos, transcends the cosmos, that he is literally *and necessarily* "not of this world." To be sure, his physical body does evidently pertain to the cosmos, the world in which we find ourselves; the point, however, is that man as such does not reduce to that physical body: *the observer or witness,* in other words, *proves to be transcendent.*

16. See *Christian Gnosis* (Angelico Press/Sophia Perennis, 2012), where I introduce the notion of anthropic realism (in Chapter 2), and show that it underlies the traditional cosmologies, from the Vedantic to that of Meister Eckhart. To be precise: anthropic realism proves to be *the only* realism tenable "in the face of gnosis."

17. "Both observer and observed," he tells us, "are part of a world that has an objective existence." (43)

Now, it happens that even from a strictly scientific point of view, the reductionist conception of the observer turns out finally to be untenable. Take the case of visual perception: in keeping with prevailing opinion, Hawking assumes that vision reduces to brain function. We are told, for example, that the human brain "reads a two-dimensional array of data from the retina and creates from it the impression of a three-dimensional space." (47) This tenet, however, has been critically challenged by an empirical scientist, named James Gibson, on the basis of experimental findings gathered in what is perhaps to date the most exhaustive inquiry into the nature of visual perception. What Gibson's experiments have brought to light is the decisive fact that perception is based, not upon a retinal image (as just about everyone had assumed), but rather on information given in the ambient optic array, which specifies, among other things, the three-dimensional structure of the environment. It appears that our visual system is designed, not simply to receive a retinal image, but to sample that ambient optic array and extract from it what Gibson terms its *invariants*. It is these invariants that are actually perceived, which is to say that *the percept is not constructed, but objectively real*: it is not simply "inside the head," but outside, as mankind had in fact always assumed. This means that what is perceived is not a visual image, be it retinal, cortical or mental, and that the so-called third dimension, in particular, is in fact no different from the other two: it too need not be constructed—by way of a process no one has yet been able, even remotely, to conceive—but is in fact directly perceived, as are all other invariants.[18]

Though widely discussed and never refuted, let me note parenthetically that Gibson's "ecological theory of visual perception" has gained no more than a partial following among cognitive scientists; and one might add, in light of considerations postponed till Part III, that acceptance of the Gibsonian paradigm by the scientific establishment at large can be effectively ruled out on other than scientific grounds. What presently concerns us, however, is the fact that Gibson's empirical findings suffice to invalidate the reductionist conception of the human observer upon which the notion of

18. For a summary and analysis of Gibson's findings see Chapter 4.

model-dependent realism is based. Take, for instance, his claim that "one can perceive an object or a whole habitat at no fixed point of observation,"[19] or that events are not perceived at a moment in time: amazing as these contentions may seem, they are simply expressive of the fact that *neither the static environment nor motion are perceived piecemeal*, as they would have to be, if perception reduced to brain function. Whatever may transpire in the brain, it is necessary, in the final stage, to bring into unity what is spatially and temporally dispersed on the level of neural activity; and this implies that *the observer, properly so called, is not himself subject to spatio-temporal bounds*. It is this transcendence of the spatio-temporal "here" and "now" that enables him to perceive "an object or a whole habitat at no fixed point of observation" as well as *movement*, something which *cannot* be detected "at an instant of time." To say, however, that the observer "transcends spatio-temporal bounds" is to declare that he is not a cosmic entity.

It happens, moreover, that substantially the same conclusion has been arrived at by way of a mathematical theorem, and curiously enough, by none other than Stephen Hawking's erstwhile mentor and collaborator, Roger Penrose. Following upon his astrophysical explorations culminating in the famous Hawking-Penrose "singularity theorem," the Oxford mathematician shifted his focus from the cosmos at large to the human brain. Neurological research had by this time established that the brain does indeed resemble a man-made computer in many respects, and the search was on to discover how this computer "made of meat" does in fact accomplish the various prodigies of human intelligence. Fixing his attention upon the solution of *mathematical* problems, in particular, Penrose asked himself whether perhaps the mathematician can solve problems that *cannot in principle* be solved by a computer, which is to say, by so-called algorithmic means. Through an ingenious application of what is commonly termed Gödel's theorem, he was able to prove

19. James Gibson, *The Ecological Theory of Visual Perception* (Hillsdale, NJ: Lawrence Erlbaum, 1986), p.197. We need to bear in mind that Gibson arrived at the claims in question, not speculatively, but on the basis of empirical evidence which in fact rules out the reductionist hypothesis.

that this is indeed the case (and one might add that the formulation and proof of this mathematical fact is itself a "non-algorithmic" accomplishment). But let us note what this entails: *it proves that human intelligence does not reduce to brain function.* Hawking's reductionist premise has thus been disproved with complete mathematical rigor.[20]

This brings us to my third point of criticism, which pertains to Hawking's ontology: his reduction of all things—all "being"—to quantum particles. Not only, thus, does Hawking reduce the observer to the status of a cosmic entity, but he goes on to reduce cosmic entities as such to "particles" which cannot be *directly* observed, cannot be *seen*; and this means that not only the observer, but the directly observable as well is ultimately reduced to brain function. Besides the fact, however, that no one has so much as the foggiest idea how the firing of a million neurons can conceivably produce such a thing as a red apple, it happens that there are weighty *scientific* grounds that militate against this hypothesis: again, the findings of James Gibson are a case in point. Philosophically speaking, Hawking's ontology reduces fundamentally to the Cartesian, which survives to this day as the hidden metaphysical premise universally presupposed by the scientific establishment at large. What stands at issue is the postulate of "bifurcation," which affirms that reality divides into an "external" world, consisting of things that can be described without residue in mathematical terms, and an "internal" world, subsisting in what Descartes calls a *res cogitans* or "thinking entity" (which Hawking identifies with the living human brain). Let us understand it clearly: *this is the undeclared ontological assumption upon which the entire edifice of Hawking's world-view is based.*

It is to be noted that this Cartesian premise cannot be tested empirically, which is to say that it cannot in principle be affirmed on scientific grounds. How, then, do we know that it is true? One might recall that Descartes himself experienced great difficulty in convincing himself that his "external" world of *res extensae*—which no

---

20. For an in-depth discussion of "neurons and mind" see Chapter 5.

human eye can ever behold—does in fact exist, and sought to justify his belief in such a world by means of a philosophic argument which appeals, finally, to "the veracity of God": the very God who has since been dismissed by crypto-Cartesian scientists, from Laplace to Hawking, as an "unnecessary hypothesis." What primarily concerns us, however, is the fact that in the twentieth century—when, according to Hawking, philosophy was at the point of death!—"bifurcation" came under rigorous attack at the hands of outstanding philosophers, beginning with Edmund Husserl and Alfred Whitehead, whose inquiries have shown the Cartesian premise to be not only unfounded but indeed untenable. Whatever else one may say regarding twentieth-century philosophy, it did, most assuredly, break the long-standing strangle-hold of the bifurcationist ontology—but only, of course, for those willing and able to heed.

The question arises now whether physics has need of the Cartesian assumption: could its findings be interpreted equally well, perhaps, in terms of a realist ontology rich enough to include what Gibson refers to as the "environment": the perceptible universe, namely, which as he notes "is not the world of physics"? It turns out that such is indeed the case;[21] and let us note, without delay, what this implies: *if it be true that the discoveries of physics can be consistently interpreted on a non-bifurcationist basis, this alone implies that it is in principle impossible to base a bifurcationist world-view upon these discoveries*, as Hawking claims to do. In the final count, the matter is as simple as that.

But there is more: as I have likewise shown in *The Quantum Enigma*, not only can physics as such be interpreted perfectly well in non-bifurcationist terms, but in fact it can *only* be "well interpreted" on that basis: for it happens that the Cartesian postulate constitutes a source of confusion and ultimately of paradox. I am referring primarily to the so-called "measurement problem"—the fact, namely, that the act of measurement interrupts the Schrödinger trajectory by "collapsing the state vector"—a phenomenon that has mystified physicists since the advent of quantum theory. Not merely, then, was

---

21. See *The Quantum Enigma* (Angelico Press/Sophia Perennis, 2012), where this question has been dealt with in all necessary detail.

Feynman right in observing that "no one understands quantum theory," but it turns out that quantum physics *cannot* in fact be understood *philosophically* in bifurcationist terms.

I will not attempt to summarize the ontological interpretation of physics enunciated in *The Quantum Enigma*. Suffice it to note that it is based upon a categorical distinction between two kinds of cosmic entities: the things that are in principle perceptible (*corporeal* objects), and those that ultimately reduce to quantum particles (*physical* objects). And this means, of course, that a corporeal object does *not* reduce to a mere aggregate of quantum particles, as almost everyone nowadays believes it does. A corporeal object, it turns out, proves to be *more* than such an aggregate; and that "more" derives from something called *substantial form*, to put it in Scholastic terms.[22] The resultant ontology—an ontology rich enough to include both the "environment" *and* "the world of physics"—differs thus from the pre-scientific through the inclusion of an additional stratum which the empiriometric enterprise of the past centuries has brought to light (or "constructed," as some believe):[23] the *physical*, namely, as distinguished from the *corporeal*. The two strata, moreover, are intimately linked (failing which physics would be impossible), and it turns out, philosophically speaking, that *the physical stands to the corporeal as potency to act*. The physical proves thus to be a *sub-corporeal* domain,[24] which is to say that measurement entails an ontological transition: a passage from potency to act.

This constitutes the key recognition, I say, which opens the door to an *ontological* understanding of quantum theory. If physics as

---

22. See pp. 31–44.

23. The view that the physical universe is in fact "constructed"—that "the mathematics is not there till we have put it there"—was first enunciated by Sir Arthur Eddington. Though never accepted by the scientific community at large, Eddington's claim has recently received strong support from the findings of Roy Frieden (see *Physics from Fisher Information*, Cambridge University Press, 1995). For an in-depth discussion of the "constructivist" view I refer to my monograph, *The Wisdom of Ancient Cosmology* (The Foundation for Traditional Studies, 2003), chap. 3.

24. Thomistically speaking, the physical domain constitutes a kind of *materia secunda* situated between prime matter and corporeal being. See *The Quantum Enigma*, op. cit., pp. 117–125.

such is indeed "the science of measurement," as Lord Kelvin observed, it follows that the so-called "measurement problem," so far from constituting a merely "technical" conundrum, refers perforce to the central mystery of quantum physics, which derives from the fact that measurement takes us *out of* the physical domain. What transpires, thus, in the act of mensuration, cannot be conceived as a *physical* process.[25] No wonder the measurement problem has proved recalcitrant to the physicist! Getting back to *The Grand Design*, I find it remarkable that an ontology that cannot comprehend the act by which physics as such is defined should have disqualified the wisdom of the ages!

Having identified the physical realm as a subcorporeal domain, we should not fail to note that this eliminates at one stroke the spectacle of "quantum weirdness," which arises from the mistaken belief that potentiae add up to make a world. The notion, for instance, that a particle moving from A to B takes every path, or that a system has, not one, but every possible history—such ideas apply precisely on the level of potentiae as distinguished from corporeal reality. By confounding these two disparate ontological domains, Hawking misleads the reader into accepting a veritable science fiction scenario. Let me be clear: what is fictional is the supposition that notions of the aforesaid kind apply to corporeal reality, that is to say, to the perceptible world, which most assuredly they do not. It needs to be understood that fundamental physics refers perforce to the *physical* world, as distinguished from the corporeal; and it is hardly surprising that mere potentiae should act in strange and unimaginable ways: there is nothing weird or paradoxical in this! It is only that Hawking has turned beautiful physics into science fiction by confounding the two ontological domains.

Hawking's ontology is Cartesian; but one should add: not quite. Like Descartes, he would reduce the objective universe to *res extensae*—to quantum particles, in his case—which necessitates that all else, all that does not reduce to quantity or mathematical structure,

---

25. The quantum description leaves out an essential ingredient, *the* essential ingredient in fact, of corporeal being, what philosophy knows as "essence." We have dealt with this question in Chapter 2.

be relegated to *res cogitans*, what Hawking terms the observer. But whereas Hawking follows Descartes in thus subjectivizing the percept, he forthwith takes a second step which the French savant was wise enough to avoid: having rid the objective universe of all that is *not* mathematical, he fills it again with a plethora of qualities by bringing the *res cogitans* back into the world of *res extensae*: "Both the observer and the observed," he tells us, "are part of a world that has objective existence." (43) Now, leaving aside the question whether this reduction of the *res cogitans* or observer to *res extensae* actually makes sense—whether it is in fact *thinkable*[26]—it happens that this step proves to be inadmissible even from a scientific point of view: this is in fact precisely what our critique of "model-dependent realism" has brought to light. But if the observer proves to be transcendent—if he does *not* reduce to quantum particles—neither do such things as red apples. By the Cartesian postulate of bifurcation—i.e., the subjectivization of the percept—all such entities have been relegated to the *res cogitans*, from whence they cannot henceforth be retrieved: Hawking can't have it both ways! If, therefore, the *res cogitans* turns out to be transcendent, so does the perceived world in its entirety. And this means that the stipulated universe of quantum particles excludes perforce not only the observer, but *ipso facto* all that is directly observed. In the memorable words of Whitehead, one is left with two things: on the side of *res extensae*, a *conjecture*, on that of *res cogitans*, a *dream*. Never mind whether the conjecture be true or false: even if it be true—even if there *is* a quantum world—there must be, in addition, something else: there is also perforce "the dreamer and his dream." So much for Hawking's ontology, which proves to be baseless and contradictory.

We come now to my fourth major point: I contend that Hawking's theory hinges upon an inadequate conception of causality. To be sure, this is hardly surprising, given what we have said earlier in

---

26. For my part, I maintain that it is *not* actually thinkable. Strictly speaking, Hawking's so-called ontology is neither science nor philosophy, but a misuse of language: a new brand of sophistry which seduces the credulous into believing the absurd.

regard to the measurement problem. It is to be noted, moreover, that when it comes to the notion of causality, Hawking himself shows signs of vacillation. At one point, thus, we are told that Laplacian determinism—the principle that "Given the state of the universe at one time, a complete set of laws fully determines both the future and the past" (30)—constitutes "the basis of all modern science," whereas forty two pages later Hawking tells us that "Given the state of a system at some time, the laws of nature determine the *probabilities* of various futures and pasts rather than determining the future and past with certainty" (72), which of course is not the same thing. My point is that Hawking is forced to hedge on this question: for as we have come to see, what finally stands at issue in "the collapse of a probability" is a passage from potency to act, something which *physical* causality cannot effect. What, then, should one say: is this "collapse of a probability" a matter simply of "chance"? Must we to suppose, in other words, that whatever has no *physical* cause, has no cause at all? I have gone to great lengths in *The Quantum Enigma* to show that such is by no means the case.

The question proves of course to be perforce philosophic: *metaphysical*, to be precise. Briefly stated, it turns out that the spatio-temporal universe—replete with its corporeal *and* physical domains— does not in fact constitute a closed system, as scientists are wont to assume. One is forced, finally, to acknowledge not only the existence of a metacosm, but of a corresponding mode of causation, which does not take place "in time"—that is to say, by way of a temporal sequence—but acts "instantaneously." In keeping with a traditional symbolism, I refer to this mode of causality as "vertical," and to the natural modes as "horizontal." What, then, are instances of *vertical* causation? In the realm of fundamental physics, as we have seen, they are precisely the acts of measurement. But there are other major realms of vertical causation, the prime example being human behavior of the kind normally associated with the idea of "free will."[27]

Consider the case of "art" in the broad sense of "human making":

---

27. *The Quantum Enigma*, op. cit., chap. 6. See also "Intelligent Design and Vertical Causation" in *The Wisdom of Ancient Cosmology*, op. cit.

can the production of an artifact be ascribed to physical or "horizontal" causation alone? I maintain that it cannot. But how can one rule out the theoretical possibility that there may indeed exist a chain of natural causation, involving billions of neurons in the artisan's brain, which does account for the production of the artifact? It happens that one can, and one can do so, in fact, with the utmost rigor by way of a mathematical theorem: I am referring to the work of William Dembski[28] which underlies what has come to be known as intelligent design or ID theory. Everyone, to be sure, recognizes instances of "intelligent design": if we come upon an assembly of stones on a hillside which spells out some message, we understand full well that it was not a rockslide that put it there; or again, if we encounter a piece of paper with a sonnet typed thereon, we know that this was not produced by a monkey banging on the keys. This raises the question whether there may be a "signature," a criterion which can perhaps be expressed in *mathematical* terms, that permits us to infer "design." Now, it is in response to this question that Dembski was led to define what he terms "complex specified information" or CSI, and prove that *no natural process, be it deterministic, random or stochastic, can produce CSI.* In our terminology, this means that CSI is a signature of vertical causation. Let me emphasize, moreover, that this is not a conjecture, a mere assertion, but indeed a mathematical fact, a *theorem.* And what does it tell us? It implies, for example, that when an artisan produces an object which carries an original[29] design (an event which entails a net increase of CSI), this artifact was *not* produced by means of horizontal causality alone: somewhere along the line an act of vertical causation *must have* entered into the causal chain. No need to know the anatomy and physiology of the brain with its myriad neurons:

---

28. *The Design Inference* (Cambridge University Press, 1998). On this subject I also refer to Chapter 10 in *The Wisdom of Ancient Cosmology,* op. cit.

29. The adjective is essential: Dembski's theorem asserts that the CSI of the object cannot be *produced* by natural causes. To invoke this result, it is therefore necessary to exclude the case where the CSI in question is supplied externally, as would be the case, for example, if the design were copied by the artisan from an external source. One sees, in particular, that the theorem does not apply in the case of mechanized or artificial production.

so long as that brain functions by the laws of physics it *cannot* account for the production of an original artifact. But this not only negates Hawking's claim that "our behavior is determined by physical laws," but disproves it *with mathematical precision*.

It is to be noted that Dembski's theory deals not simply with "design," but indeed with "intelligent design": what does this mean? It appears that Roger Penrose, in his study of what computers or brains can and cannot do, has hit upon the answer when he concludes that "the very essence of consciousness" consists in an internal "seeing," an "ability to divine (or intuit) truth from falsity (and beauty from ugliness!) in appropriate circumstances."[30] Whether it be a question of judgments that cannot be formed by algorithmic means or of acts productive of CSI, what counts is a certain "seeing," an intellective apprehension, be it of truth or of beauty (and if beauty be indeed "the splendor of truth" as Plato declares, the two are closely knit). It follows, now, that at the very core of a human being, intelligence or "intellect" comes into play: something which does not reduce to brain function and enables acts which physical causality cannot effect.

But for Hawking there is only physical causality and its absence, called "chance," which supposedly explains why a probability distribution collapses for no assignable reason at all. One *sees* now (here it is again, this word!) that if such were the case, one would have to conclude, among a host of other absurdities, that all non-algorithmic judgments—including those which underlie Hawking's own doctrine—are reached "by chance," which would of course imply that they carry no weight at all. Generally speaking, the denial of vertical causality in the human domain entails the denial of intelligence, and constitutes therefore a *reductio ad absurdum* of the aforesaid denial itself. It is needless to say more.

This brings us finally to Hawking's stand on "creation." From the outset he attacks the notion of a Creator, and hardly misses an opportunity to deride the belief that a beneficent God "*created the heavens and the earth*." He argues that such an hypothesis is both

---

30. *The Emperor's New Mind*, op. cit., p. 412.

unnecessary and unreasonable, that in fact a virtually infinite number of universes emerge from a preceding vacuum through the operation of physical laws, which is to say that creation reduces ultimately to "a quantum event."

To begin with, it behooves us to note that the idea of "creation" proves to be incurably *metaphysical*. We need to realize, in the first place, that the creative Act cannot be conceived in *temporal* terms: *creation* is not something that happens "in time." As Meister Eckhart states with the utmost clarity: "God makes the world and all things in this present now," that is to say, in the *nunc stans* or "now" which is *not* a moment in time. And what is it that "God makes"? We are told that it is "the world *and all things*." Now, a great deal of confusion has ensued, even occasionally in theological discourse, because one tends to forget the second part of this assertion. There are those who think that whereas God brought the world into existence ages ago, it has been running on its own ever since; but this notion is doubly mistaken: first, because it places the creative Act in the past; and secondly, in that it reduces what God has made to a mere initial state. Admittedly, the idea that "God makes the world and all things in this present now" is hard to comprehend, and in fact entails the difficult conception of an aeviternal metacosm or primary world; but this simply implies that whosoever wishes to challenge that immemorial doctrine must take care not to refute an *Ersatz*.

Yet this is precisely what Hawking's argument is designed to accomplish, whether he knows it or not. His strategy is to depict the Judeo-Christian doctrine as a kind of primitive science, a "model" designed to explain observable facts. It is actually all he *can* do; for so long as the doctrine is conceived on its own level—that is to say, in authentically metaphysical terms—it is *ipso facto* immune from attack on strictly physical grounds. Physics is not, after all, equipped to speak of metacosmic realities: from its own point of vantage, such notions are judged perforce to be "meaningless." To make his case on *physical* grounds, Hawking requires a corresponding criterion by which the perennial doctrine is to be judged.

He broaches the subject of "creation" with a famous Augustinian dictum: "The world was made, not *in* time, but *with* time" (50), which he respects as legitimate in its own right. "This is one possible model," he tells us. Now, everything hinges upon that word, "model," which has obviously been chosen on account of its *scientific* connotations. By way of this inappropriate and misleading designation—this semantic trick!—Hawking depicts the metaphysical doctrine of "creation" as a kind of primitive physics, a rudimentary science which as such can be compared with *our* science. We are left with two competing "models": the Biblical and that of 21st century physics. His straw man thus emplaced, Hawking forthwith makes his point: "The second model," he informs us, "can explain the fossil and radioactive records and the fact that we receive light from galaxies millions of light-years from us...," all of which, needless to say, the first "model" can *not* do. But even if one grants that "radioactive records" and "galaxies millions of light-years from us" are indeed factual, and can be explained by means of contemporary physics, this in itself would hardly disqualify the contention that "God created the world"—*unless*, of course, that tenet has first been reduced to the status of a "competing model."

Such is the reductionist contention Hawking brings into play regarding the nature and function of creationist doctrine; and to be sure, he does so surreptitiously, and without a shred of evidence in support of that claim. We need not detain ourselves further with this baseless hypothesis; it will however be of interest to say a few words concerning the "explanatory value" of the metaphysical doctrine Hawking wishes to disqualify. It is to be noted, first of all, that the creative Act is evidently "causal" in the extreme, inasmuch as it brings into existence "the world and all things." But that causality, if one may call it such, proves to be "vertical" in that it is evidently not mediated by a temporal sequence of events. What is more, the creative Act constitutes in fact *the prototype and principle of all vertical causation*, which is to say that vertical causality, properly so called, constitutes a secondary mode of creation, a kind mediated by created agents. What stands at issue is *the miracle of intelligence*, which is precisely what distinguishes vertical from horizontal causality. Certainly there are different kinds of intelligent mediation, ranging from the angelic—which is not, after all, a figment of the primitive imagination!—to the human, that give rise to corresponding modes of vertical causation.[31] The point, once again, is that an intelligent agent, no less than a so-called observer, does not reduce simply to a cosmic entity. It needs finally to be recognized that intelligence hinges upon a transcendent reality, something that is in fact divine: "*the true Light*," namely, "*which lighteth every man that cometh into the world*."

But the question remains: can there be a *science* based upon vertical causality, even as there are sciences based upon physical causation? Now, it happens that there can, and that such sciences have in fact existed since ancient times:[32] traditional or "sacred" sciences, one may call them; it is only that *our* sciences, geared as they are to *physical* causality, are in principle incapable of understanding a science based upon vertical causation. The traditional sciences, to be sure, have their own modus operandi, which needless to say, differs radically from the empiriometric. So too they have an "explanatory value" and usefulness of their own, which do not, to say the least, compare unfavorably with the benefits to be derived from the physical sciences of our day.[33] Of course this is hardly the place to dwell more deeply on the traditional sciences and their relation to the physical; I wish only to make one further point: namely, that these two *kinds* of science do not stand in conflict or contradiction, that it is not a question of "either or." As I have shown elsewhere,[34] horizontal and vertical modes of causality can and do coexist without interference, which is to say that each has its own proper effect. Take a simple example: a marksman fires at a target. Now, from the standpoint of horizontal causality, the ensuing impact is explained in terms of a temporal sequence of events initiated by the pull of a trigger, whereas the same effect is equally the result of an intentional act: neither explanation disqualifies the other, and of course there can be no question as to which is "more true."

31. *The Wisdom of Ancient Cosmology*, op. cit., pp. 194–198.

32. The only traditional science to survive, in the West, to the present day appears to be astrology. Its sister science, alchemy, has all but disappeared in the course of the seventeenth century.

33. For a glimpse of what traditional science can accomplish—the stupendous scope, accuracy, and explanatory value it can achieve—see Chapter 6.

34. *The Quantum Enigma*, op. cit., pp. 114–17.

But Hawking evidently has no inkling that there *are* sciences other the contemporary, let alone that the two kinds are not opposed but complementary: his inability to recognize the existence of vertical causation predisposes him to judge the worth of all doctrine in terms of its capacity to explain phenomena by way of the one and only causality he knows: the horizontal mode, as conceived by the physicist.

Getting back to Hawking's argument: It now appears that God— the Creator of *"the heavens and the earth"* —has indeed survived the attack; once the smoke has cleared, one sees that his straw man argument carries no weight at all. *But that is only half of the story: for not only his argument against the doctrine of a divine Creator, but Hawking's own version of cosmogenesis—which is supposed to replace the Judeo-Christian teaching—is fatally flawed as well.* Consider the previously noted fact that the physical universe proves *not* to be a closed system after all, which is to say, once again, that vertical causation comes perforce into play. As we have pointed out under the rubric of "causality," it comes into play in every quantum-physical measurement, as also in every act based upon human intelligence, beginning with the production of an artifact. But Hawking would have us believe that contemporary physics is able, in principle, not only to explain the functioning of the observable universe, inclusive of man, but to disclose in addition how that universe came to be. Now, this latter claim seems strange, given the fact that even after the universe is in place, entities emerge which *demonstrably* cannot be produced by way of physical causation. If physical causes prove to be incapable of producing even a water-pot from pre-existing clay, one wonders how these same causes could give rise to the universe at large! And in fact, they can not: for in asserting that the universe itself has been brought into existence by physical causes, Hawking affirms that what has thus been brought into existence is not simply an initial state, but includes perforce whatever exists or transpires in that universe. It follows that *a single demonstrable act of vertical causation suffices to disqualify Hawking's thesis.*

We need not belabor the point. As one should have surmised from the start, the claims Hawking puts forth on the subject of "creation" prove in the end to be unsubstantiated and untenable. Not only has contemporary physics *not* in truth disproved the authentic tenets of creationist doctrine, but it turns out that the defamed doctrine is ultimately needed to understand physics itself: what it can and cannot do. In the final count, *bona fide* metaphysical conceptions do perforce enter the picture, whether the scientist likes it or not, for the simple reason that both the universe and the vertical causation operative therein derive from a transcendent reality concerning which physical science as such knows nothing at all.

III

It behooves us, finally, to put Hawking's claims in perspective by taking a closer look at the contemporary scientific enterprise as such. We need to transcend what we have been taught in schools and universities to discover on our own what we are never told: only thus can we begin to perceive the full picture. To place *The Grand Design* within the context of the existing culture, it is above all imperative to get over the notion that science is simply a quest in search of truth: open, unbiased and fair. We need to realize that the enterprise has an ideology, an agenda, an establishment, and vested interests to protect; as anyone past childhood should realize, "politics" does enter the picture.

In line with these general observations I would like to point out that Hawking overstates the *scientific* case in support of his claims by suppressing all contrary evidence. He does so most blatantly, to be sure, in his treatment of the Darwinist theory, which evidently constitutes a necessary component of his world-view: nowhere does he give even the slightest indication that there yet remain unresolved questions of a basic kind, let alone acknowledge the fact that all sorts of adverse evidence has been piling up for more than a century, and that, from a strictly scientific point of view, the Darwinist hypothesis should have been rejected long ago.[35] Even the publica-

35. The serious anti-Darwinist literature of our day, though largely ignored by the media, has witnessed a spectacular growth in recent decades. To mention at least a few representative titles: Michael J. Behe, *Darwin's Black Box: The Bio-Chemical Challenge to Evolution* (NY: Free Press, 1996); William A. Dembski, *The Design Revolution* (Downers Grove, IL: Inter-Varsity Press, 2004); Michael Denton,

tion of William Dembski's theorem—which shows that evolution *à la* Darwin proves to be impossible on strictly mathematical grounds—seems to have had no effect on Hawking: he continues blithely to treat Darwinian evolution as a scientifically established fact.

We need now to ask ourselves how the scientific case stands when it comes to *physical* theories such as Einsteinian relativity and big bang cosmology: have these perhaps been rigorously verified beyond reasonable doubt? Admittedly, this is a difficult and perforce technical question; yet I propose to shed light on the issue by showing that even here, in this rarefied technical domain, an element of *ideology* does of necessity come into play. It does so, moreover, not simply as a syndrome of beliefs and values which impel the scientist to pursue his inquiry, or define the direction of this quest, but indeed as a determinant of the resultant theory, of what in the end is found or discovered. Simply put, I maintain that *the world-view at which science arrives by purportedly rigorous means proves finally to be reflective of the ideological assumptions that guided the enterprise from the start.*

Let me begin by recalling an event: When in 1965 Arno Penzias and Robert Wilson picked up signals from outer space said to derive from the microwave background, the *New York Times* announced the discovery with the headline: "SIGNALS IMPLY A BIG-BANG UNIVERSE." By way of contrast, let me now recall what happened in 1887, when Albert Michelson and Edward Morley conducted their experiment designed to measure the velocity of the Earth in its orbital motion around the sun. What they found—to the dismay of the scientific community!—was that this velocity, so far from being the expected 30 or so kilometers per second, turns out to be, precisely, *zero.* And let us note that this result was in no way uncertain or tenuous: based upon the laws of what is nowadays, in retrospect, termed "classical" physics, the fact that *the Earth does not move* was strictly implied by the outcome of the experiment. But whereas this finding sent shock waves through the scientific world, the public at large was told little. Most assuredly, there was no banner headline proclaiming that "MEASUREMENTS IMPLY AN IMMOBILE EARTH," which unlike the 1965 caption, would not have been a mere journalistic exaggeration, but indeed a scientifically accurate statement.

What eventually happened, in response to the Michelson-Morley finding, is the advent of a new physics, consisting of the special and general theories of relativity, which gets around this *ideologically* unwelcome result through the stipulation that the observed speed of light is the same in all so-called "inertial" frames of reference. And needless to say, this event did receive its full share of publicity: as everyone knows, Albert Einstein, almost overnight, became a scientific superstar, and his theory of relativity a scientific breakthrough of the first magnitude. But the question remains: *is it true?* Does Einsteinian physics square, actually and fully, with the observable facts (at least in situations where quantum effects may be neglected), as Hawking, and indeed the scientific establishment at large, aver? My point is that this question proves to be far more difficult than one is led to suppose: as in the case of Darwinism, the matter is by no means as clear-cut as Hawking would have us believe. Only one thing is certain: the choice lies between geocentrism and Einstein.[36]

Having identified "the constancy of the speed of light" as an ideologically motivated postulate (verified or not, as the case may be), I would like now to point out a second ideological premise which likewise proves essential to Hawking's world-view. What stands at issue,

36. On this issue mention should be made of a remarkable 2-volume treatise by Robert A. Sungenis and Robert J. Bennett, entitled *Galileo Was Wrong* (the fifth edition of which was published in 2008), a work which constitutes presumably the most exhaustive study of this question to date. The book covers over eleven hundred folio pages and gives well over a thousand references, a good part of which derive from the scientific journal literature, in support of the contention that Einsteinian physics has been *de jure* disqualified. But whereas much of what the authors bring to light is indeed cogent and does bear adversely upon the Einsteinian claims, the work as a whole is unfortunately marred by an excessive polemic which at times overshoots the mark.

this time, are not the laws of physics, but the structure of the universe as conceived in astrophysical cosmology. This too, it turns out, hinges upon an ideological postulate; and strangely enough, it is Hawking himself who tells us so in an earlier treatise: "We are not able to make cosmological models," he writes, "without some admixture of ideology."[37] What he is referring to, in particular, is the assumption that stellar matter, when viewed on a sufficiently large scale, is uniformly distributed throughout space (much like the molecules in a gas, which appears to have a uniform mass distribution given by a density). Now, this is an assumption, an *ideological postulate* no less, as Hawking himself informs us. But what is it that renders the premise "ideological"? This too Hawking explains: "We shall, following Bondi, call this assumption the *Copernican principle*," he goes on to say. Here we have it: what stands at issue, once again, is a repudiation of "geocentrism" in the wide sense of a cosmic architecture reflective of intelligence—of *intelligent design*, that is— and thus of *an intelligent Creator*.

Think of it: here Hawking himself is telling us that this repudiation or denial of design on a cosmic scale is *not* in fact a scientific discovery—a reasoned conclusion based upon observable facts— but constitutes rather "an admixture of ideology"! Yet surprising as this admission may seem in light of what we have been taught to believe, it is easy enough to recognize that it is impossible to base a cosmology upon strictly scientific ground. The fundamental problem is this: since one is unable, in the astrophysical domain, to act upon the source of the signals received, one cannot carry out the kind of controlled experiments upon which physics as such is based. One may put it this way: whereas physics deals finally with what John Wheeler terms a "participatory universe," the cosmos at large happens *not* to be participatory. Astrophysical cosmology, therefore, *is not physics*: it cannot be. And as a matter of fact, that cosmology does rest upon an "ideological postulate" in the form of the Copernican principle, as Hawking himself informs us. In plain terms, we are told that *the a priori denial of intelligent design on a cosmic scale*

---

37. *The Large-Scale Structure of Space-time* (Cambridge University Press, 1973), p. 134.

constitutes the ideological assumption upon which big bang cosmology is based. To which one should add that Hawking's argument against "the Grand Designer" proves thus to be circular, and that physics, properly so called, can establish no such thing.

There remains the question of evidential basis, of verification. It is to be noted, first of all, that in the absence of controlled experiment, verification in the full scientific sense is ruled out in advance: the best one can hope for is that signals from outer space, when interpreted according to terrestrial physics, do not conflict with the theory. It happens, however, that they do, which is to say that it has been necessary to introduce a number of ad hoc hypotheses: i.e., assumptions formulated specifically for the purpose of squaring the theory with conflicting observational findings.[38] What is more, the process of adding extra assumptions in response to adverse data appears to be ongoing; as Brent Tully (known for his discovery of supergalaxies) observed: "It's disturbing that there is a new theory every time there is a new observation." To which one might add that Tully has every right to be disturbed: for such a modus operandi in effect eliminates empirical verification as a criterion of truth. Under such auspices it becomes hard to ascertain whether there exists so much as a shred of *real* evidence in support of the theory.

Yet Hawking has not one word to say on that score: we are given to believe that big bang cosmology is simply physics, and as such has been rigorously substantiated, once and for all, on unimpeachable scientific grounds. The need for "an admixture of ideology," in particular, is nowhere mentioned in *The Grand Design*: on the contrary, Hawking makes it a point to convey the impression that "M-theory" alone—the ultimate science!—guarantees everything he has to say.

A certain similarity between big bang cosmology and Darwinism has thus come into view, an analogy which it may be enlightening to reflect upon. No less than the astrophysical cosmology, Darwinist biology is a reputedly scientific theory advanced on insufficient grounds, which is to say that both are in reality advanced on *ideological* grounds. It needs moreover to be recognized that the respec-

---

38. I have touched upon these questions in *The Wisdom of Ancient Cosmology,* op. cit., chap. 7.

tive theories spring in fact from one and the same ideological postulate: whether it be a question of species or of the universe at large, *evolution*—the negation of *intelligent design!*—proves to be the founding dogma of the one as of the other. In a word, *big bang cosmology is Darwinism on a cosmic scale.* And needless to say, this fact does prove to be enlightening, all the more since at present the biological Darwinism is understood far better than the astrophysical.[39] The salient point which emerges with special clarity in the biological domain is that *Darwinism is never science*; no matter what garb it dons, it remains in essence what it was from the start: *an ideology.* And this means that "evidence" loses its primacy: it is still desirable, still sought after, but ceases to be necessary, inasmuch as the theory stands ultimately on ideological ground. One is reminded of the Darwinist Ernest Mayr's response when confronted by calculations demonstrating the astronomical improbability of the evolutionist scenario in the case of a human eye: "Somehow or other, by adjusting these figures," said he, "we will come out all right. We are comforted by the fact that evolution has occurred."[40] The point has meanwhile been made with utmost clarity by Richard Lewontin, himself a leading evolutionary biologist; commenting on science at large, he writes:

> We take the side of science in spite of the patent absurdity of some of its constructs, in spite of its failure to fulfill some of its extravagant promises for health and life, in spite of the toleration of the scientific community for unsubstantiated just-so stories, because we have a prior commitment to materialism. It is not that the methods and institutions of science compel us to accept a material explanation of the phenomenal world, but on the contrary, that we are forced by our *a priori* adherence to material causes to create an apparatus of investigation and a set of concepts that produce material explanations, no matter how counter-intuitive, no matter how mystifying to the unini-

---

39. Mention has already been made of the growing literature which has in effect "unmasked" the biological Darwinism. See fn. 34 for a few representative titles.

40. Quoted by Phillip Johnson in *Darwin on Trial*, op. cit., p. 38.

tiated. Moreover, that materialism is absolute, for we cannot allow a Divine Foot in the door.[41]

One more point needs to be made: the case of science *properly so called* is different. When it comes to *fundamental* physics, in particular—which is and can be none other than quantum theory— what confronts us is indeed the authentic "science of measurement." Yes, ideology did no doubt motivate the founders—from Bohr to Heisenberg, Schrödinger and Feynman—and direct their focus towards the quantitative pole of cosmic manifestation;[42] yet it did not interfere with the legitimate modus operandi of a mathematical physics: it did not *force* the result. As a matter of fact, the very opposite is true: insofar as quantum mechanics contradicts the long-standing canon of Laplacian determinism, its discovery was profoundly distasteful to the physics community at large, as Hawking himself points out. It needs thus to be noted that quantum physics most certainly did *not* commend itself on ideological grounds, but imposed itself, rather, on the basis of irrefutable empirical evidence. For more than eight decades, moreover, it has continued to distinguish itself by the unprecedented scope and uncanny accuracy of its predictions: in a thousand experiments it has never yet proved wrong. No need, in this domain, for ad hoc hypotheses: the inner logic of quantum theory itself, interacting with experimental findings, drives the development. Leaving aside the penumbra of scientistic notions which surround the discipline without corrupting it, what confronts us here constitutes, quite evidently, the most brilliant and spectacularly successful achievement of physical science as such. What a pity that Hawking has spoiled *beautiful physics* with baseless and amateurish speculations of a pseudo-philosophic kind!

41. *The New York Times Review of Books*, 9 January, 1997. Quoted by Bruce L. Gordon in "Balloons on a String: A Critique of Multiverse Cosmology" (*The Nature of Nature*, B.L. Gordon and W.A. Dembski, eds., Intercollegiate Studies Institute, 2011), p. 584.

42. See pp. 31–41.

8

METAPHYSICS AS "SEEING"

SINCE THE BEGINNING of modern times, metaphysics has been viewed as an academic discipline, to be pursued at universities; and it is of interest to note that, as such, its standing and prestige in the educated world has steadily declined, to the point where many nowadays deny its philosophic legitimacy. Yet I contend that the metaphysical quest pertains by right, not to the artificial environment of the contemporary university, but to human life, human existence in its untruncated reality. In plain words: it springs from man's innate thirst for *truth*, which is none other than the thirst for God, who nowadays "is not mentioned in polite society" as Ananda Coomaraswamy reminds us. Metaphysics is therefore something that concerns each of us by virtue of the fact that we are human, which is to say, "made in the image and likeness of God." It is indeed a case of "*noblesse oblige*": so far from reducing to a mere academic discipline—to be pursued by "professionals," notably recipients of a doctorate in philosophy—metaphysics constitutes an activity of the mind and heart to which, in principle, all are not only entitled, but are, in a way, "*called.*"

It is to be noted that our preconceived notions regarding metaphysics tend generally to be not just inaccurate, but in a way inverted or "upside-down." We are prone, first of all, to imagine that the discipline stems from "doubt," when in fact it springs from a profound sense of "wonder," which is actually the very opposite of doubt: for that wonder proves to be in essence a recognition, however dim, of the inscrutable immanence of God in the things of this world. So too we tend to think that the means or *modus operandi* of metaphysics consists of reasoning, that is to say, of rational argument, when in fact it is, again, the very opposite: a question, namely, of "seeing," of direct perception, of *gnosis* properly so called. Admittedly, reasoning does have a role to play; but its function is inherently negative and preparatory; to be precise, rational argument serves to deconstruct false beliefs, and in so doing, to *purify the mind*. That is all it *can* do, and indeed all it needs to do; for to the extent that the mind has been purified—the "mirror" wiped clean—the "seeing" takes care of itself. This holds true to the very end: as the Savior assures us: "the pure in heart shall see God."

We need however to realize that God enters the picture, not only at the end of the metaphysical quest, but from the very outset, and not only as *object* of the aforesaid "wonder," but in a way as its *subject* as well. Indeed, we could in no wise "sense" God outside of ourselves if He were not also present within the depths of our soul as the first and ultimate "seer." It is this inscrutable indwelling of God—as the "soul of our soul"—that enables and indeed powers the quest from its first inception to its ultimate end. We need thus to divest ourselves of the idea that the metaphysician is simply "this-man-so-and-so": if such were the case, the enterprise could never succeed, nor in fact could it even begin. It may have been this realization that prompted the aged Husserl—one of the greatest philosophers of the twentieth century—to confide sadly, one day, to Edith Stein (his former disciple, who by then had become a Carmelite nun on her way to sainthood): "I tried to find God without God!"

We have maintained, in keeping with sapiential tradition, that metaphysics is inherently a "seeing"; it needs also, however, to be noted that every "seeing"—even the humblest act of sense perception—is in a way metaphysical, and can in principle serve to initiate the metaphysical quest. It is a question of following what may be termed "the spoor of God" in visible things: "For the invisible things of Him from the creation of the world are clearly seen, being understood from the things that are made."[1] One may take this to mean that what St. Paul refers to as "the invisible things of God" *are* in fact what is "clearly seen," which is to say that they are precisely what *would be* seen, if indeed we saw "clearly." St. Paul is putting us on notice that in "seeing" we generally "see not." We are

---

1. Romans 1:20.

given to understand that a collective blindness has overtaken us, which the Apostle goes on to ascribe to an apostasy, an estrangement from God: "Because that, when they knew God, they glorified Him not as God, neither were thankful; but became vain in their imaginations, and their foolish heart was darkened."[2] Notwithstanding our customary belief in "progress" and progressive enlightenment, it happens that Christianity teaches the very opposite: it affirms that there has been not only a primordial Fall, but indeed an ongoing decline in our ability to *see*. We seem in fact, in this postmodernist age, to be nearing the final stage of that collective deterioration, a condition which St. Paul proceeds to characterize: "Professing themselves to be wise," he declares, "they became fools."[3] Now, unflattering as this depiction may be, it behooves us to take the Apostle at his word. The primary task of the true metaphysician is then to undo that collective decline, to reverse it in himself. It is a question of restoring the "heart" from its "darkened" condition, and in so doing, to recover the unimpaired use of our God-given "eyes": such, in brief, is the task of veritable metaphysics. We need not theorize as to who precisely the metaphysician himself may be, nor what will become of him when his heart has been "undarkened": that is something which shall remain a mystery until the work is done. As St. John the Evangelist informs us: "It doth not yet appear what we shall be."[4]

To comprehend what it actually means "to see," one needs, first of all, to divest oneself of the Cartesian dualism which our education has imposed upon us, whether we realize it or not. This imperious philosophy reduces to the supposition that integral reality splits neatly into two domains: an objective world, comprised of "extended entities," and a subjective realm, made up of so-called "consciousness." It happens, however, that this dichotomy is ill-founded and indeed spurious, a fact which not only accords with

2. Ibid., 1:21.
3. Ibid., 1:22.
4. 1 John 3:2.

the great metaphysical teachings of antiquity, but, as it happens, has been recognized by leading philosophers of the twentieth century, beginning with Edmund Husserl and his erstwhile follower, Martin Heidegger. The point is that "seeing" does not reduce to the "reception into consciousness" of something that pre-exists in the external world, but constitutes rather an "act of intentionality" which conditions and in a way "defines" its object. What is more, consciousness is not something which precedes that "act," but *is itself* that act, which is to say that it is never without content—like an empty receptacle—but is invariably a "consciousness *of.*" So too, what antecedes the intentional act "externally" is not in fact the object or "extended entity," but the *phenomenon*, conceived (according to the literal sense of that Greek word) as "that which shows itself in itself." It is to be noted, moreover, that the phenomenon, by virtue of the fact that it shows itself "*in itself*"—that is to say, not just in some representation, some private phantasm, but literally "in itself"— does not belong "exclusively" to the external or objective side of the Cartesian divide: it breaks the dichotomy, in other words. To be sure, given the contemporary bias, it is not surprising that the word should have lost its original sense, and has in fact come to mean virtually the opposite: an effect or manifestation, namely, of a reality which stands forever "behind" the phenomenon. To put it in standard Cartesian terms: the "real" consists supposedly of *res extensae* or "extended things," situated in the external world, while the "phenomenon" has been reduced, in effect, to a subjective apparition, contained within what Descartes terms a *res cogitans* or "thinking entity." All qualities, in particular, beginning with colors— everything, in other words, that cannot be conceived in quantitative or mathematical terms—has been excluded from the real or "external" half of the Cartesian divide and relegated to the *res cogitans*. What, then, does it mean "to see"? It means perforce to behold a private apparition pertaining to one's own *res cogitans*.

One may, of course, ask on what grounds these stupendous conclusions have been reached: what, in other words, is the evidence— be it empirical or *a priori*—in support of the Cartesian premises? Suffice it to say that there is in fact no evidence at all: these very postulates preclude that there *can be*. And yet, strange to say, the premises in question have dominated our so-called "scientific" worldview from the start, and continue to do so to the present day. Now as before it is the official credo of science that each of us is cooped up in his own "consciousness," his own *res cogitans*, constrained to gaze, without reprieve, upon apparitions generated somehow by external causes. We have all, of course, learned to live with this impasse: it is what our so-called "higher" education has obliged us to do. It should however be noted that in fact—mercifully!—not a single human being accepts this Cartesian stipulation in his or her daily life: to do so would constitute insanity.[5] Instead, we have learned to oscillate, as it were, between our "daily" *Weltanschauung* and the Cartesian—which we uphold in our scientific convictions—without so much as realizing that these two orientations stand in stark contradiction: that one moment the grass is green, and the next it is not!

Meanwhile something altogether unexpected has come to pass, which we should at least touch upon: in the early decades of the twentieth century—the very period in which Husserl and others came to recognize the philosophic absurdity of the Cartesian claim—it happened that physics itself has in a way disavowed that philosophy. This is not to say, to be sure, that physicists *en masse* have abandoned these philosophic assumptions: nothing, in fact, could be further from the truth. What has happened, rather, is that with the discovery of quantum mechanics (around 1926), physics could no longer be interpreted in Cartesian terms, which is to say that certain quantum-mechanical findings—most especially the so-called "state vector collapse"—assumed the appearance of outright paradox. Now, it can be shown that the paradox disappears the moment one abandons the Cartesian premises, that is to say, the hypothetical dichotomy of "extended things" versus "*res cogitans*."[6]

---

5. However, despite this pervasive disbelief, the Cartesian doctrine has had a profound effect on the Western psyche, to the point of provoking a kind of "collective schizophrenia," a matter with which I have dealt at length in *Cosmos and Transcendence* (Tacoma, WA: Angelico Press/Sophia Perennis, 2012).

6. See my treatise, *The Quantum Enigma* (Tacoma, WA: Angelico Press/Sophia Perennis, 2012).

It turns out that "quantum paradox" is simply Nature's way of repudiating a spurious philosophy.

One finds in the end that the real proves to be what indeed it *must*, namely "that which shows itself in itself"; in a word, it is in fact the *phenomenon* in precisely the original sense of that term. Oddly enough, however, what we ordinarily perceive is something else! On the strength of Husserl's "phenomenological" analysis one may conclude that the "seen" falls short of the phenomenon, that in fact it is subsequent to the "seeing," which may be said to break up into a *perceived* object and a *perceiving* subject. The two constitute thus a complementarity: the perceived object and the perceiving subject belong together, like the faces of a coin. What "precedes" that complementarity—what is primary—is the intentional act itself, that is to say, the actual "seeing." The act comes first, and "by the time" the separated subject and its "external" object present themselves, the actual "seeing" has come to an end; as Henri Bortoft[7] has keenly put it: "We are always too late!" For indeed, in the actual "seeing," subject and object are *not* separated: as Aristotle has observed, "in a certain manner" the two are "one."

Husserl understands, in his own way, that "seeing, they see not," as Christ declared to the multitude[8]: for him the "not seeing" results from the break-up of the intentional act, the fact that "we are always too late." To overcome this failing, this congenital blindness, we need evidently to seize the intentional act at an "earlier" moment, so to speak, "before" it breaks into the familiar subject and its concomitant object. That "before," however, proves not to be temporal, but "ontological," if one may put it so; it has to do, not with temporal sequence, but with levels of awareness. In a word, "before" means "deeper," or as one can also say, "more primary." We need not concern ourselves with the technical vocabulary Husserl devised in an effort to communicate to the philosophic community at large what he had discovered or brought to light; suffice it to say that his method entails a

---

7. A theoretical physicist (and student of David Bohm, no less), Bortoft belongs to that exceedingly small contingent of present-day scientists who have transcended the contemporary scientistic world-view.

8. Matt. 13:13.

"standing back" from the familiar perceptual act, as if to observe that act from a deeper ground. Husserl's *modus operandi* was in a way the opposite of what philosophers are wont to do: instead of conceptualizing, he "*deconceptualized*" in order to "*see.*" One may think of him as a "philosophical archeologist," seeking to expose deeper levels of awareness by clearing away layer upon layer of mental constructs, by which these "earlier" strata had come to be overlaid.

It is to be noted that in this regard, at least, Husserl's approach is kindred to that of the great sapiential traditions, which likewise recognize "deeper" levels of perception and entail a hierarchic conception of the percipient. Briefly stated, the veritable *anthropos* is held to possess not only a periphery (where our conscious acts "normally" take place), but an absolute center as well, and is said to comprise, in addition, a hierarchy of intermediary "centers," each of which defines a "level of vision" and a corresponding state.[9] What, then, does it mean "to see"? In the final count, it means to perceive from the deepest center of all, sometimes termed the "heart"; and that is indeed the primary and authentic "seeing," from which mankind has become progressively alienated, starting with the Fall.

Beginning from where we presently stand, let us now ask what it is that "precedes" the "external" object: what does he see who is *not* "too late"? In the terms of Husserl's analysis one is bound to reply that it is precisely the phenomenon, conceived as "that which shows itself in itself." But then, what *is* it that thus "shows itself"? Of course one may reply that this is something everyone will have to discover for himself by applying the appropriate means, a necessity which can be neither denied nor circumvented. Yet, even so, there is something to be learned from the testimony of those who *have* pursued that path, be they philosophers, poets, artists, or mystics of some kind.

---

9. The fullest description of these "centers" has no doubt been given in the Tantric tradition of India, which refers to them as "*cakras*" (literally "wheels") and "*padmas*" ("lotuses"). Kashmiri Tantrism, in particular, has evolved a veritable science concerned with this subject. See Chapter 6, pp. 135–137.

The field, quite clearly, is vast. What I now propose to do, by way of selection, is to highlight the *scientific* ideas of a man known mainly as a poet and literary figure, who spoke with consummate precision on the subject at hand, at a time when there was hardly anyone to pay heed. As the reader may have surmised, I am referring to none other than Johann Wolfgang Goethe, whose conception of "science" was radically opposed to the Newtonian, and reduces in fact to what he terms "*Anschauung*": a "seeing" of the phenomenon itself. "Don't look for anything behind the phenomena" he tells us; "they themselves are the theory." They *are* "the theory"—not, of course, in the sense of an abstract conception, let alone of a mathematical formula which supposedly describes a reality no one perceives—but rather in the original sense of "*theoria*": as an actual "seeing," a direct knowing in which, "in a certain manner," subject and object do indeed "become one." This is what it takes *not* to be "too late"! One must not, however, think of Goethe as a so-called "Platonist": for the German poet and artist, "knowing" was not "merely intellective," but entails actual vision, the kind that involves our corporeal eyes. The Goethean "*Anschauung*" is neither purely intellective nor exclusively sensuous, but may be characterized as "an intuitive knowledge gained through contemplation of the *visible* aspect," as someone has aptly said. It constitutes a "seeing," thus, in which the subject, so far from being merely a recipient of something given, is an active participant. To be precise, the "seer" is called upon to *penetrate* the "visible aspect" which he receives, and in so doing, lay hold of its very essence: in a word, authentic seeing constitutes an *act*.

The first point to be noted, regarding knowledge thus derived, is that its object is not a sum of parts, but is perforce a whole; as Bortoft explains: the object of such knowing is "the whole which is no-thing," and tends consequently to be mistaken for a mere "nothing," in which case it vanishes. "When this happens, we are left with a world of things, and the apparent task of putting them together to make a whole. Such an effort disregards the authentic whole."[10] It is

10. *The Wholeness of Nature* (Hudson, NY: Lindisfarne Press, 1996), p. 14. This is no doubt the best book on "Goethe's Way toward a Science of Conscious Participation in Nature," as the subtitle has it.

at this point, clearly, that "science" in the Baconian sense—and modern physics, in particular—comes into play: as is the case with our seeing, our science likewise is "always too late." Unable to deal with the veritable phenomenon directly—through our God-given eyes—the Baconian scientist postulates a mechanism (or, from an operational point of view, a "model" of some kind) to explain what he *can* see, what Bortoft calls "a world of things." It is well known that Goethe opposed this approach, that he absolutely rejected the idea of a mechanism "behind" the phenomenon; but the question is: upon what grounds? One may answer as follows. He understood, first of all, that the real is in fact none other than that which *can* in principle be known, and that this is ultimately the phenomenon: "that which shows itself in itself." And he realized, further, that the authentic phenomenon is indeed "the whole that is no-thing," as Bortoft points out. It remains now to observe that this "no-thing" cannot be a mechanism, because it is not a sum of parts. Such seems to be the unspoken argument or implicit train of thought by which Goethe arrived at the realization which shocked his contemporaries: his categorical denial, namely, of Newtonian mechanism. In his ongoing dispute with the scientific authorities of the day, he insists, time and again, that *there is no mechanism*: nothing in fact that stands "behind" what is in the deepest sense "seeable." There cannot be, because, in the final count—to say it again—the real *is* the seeable: "that which shows itself in itself."

I would like now to point out that the Goethean "denial of mechanism"—which in his day was met with derision, bordering upon contempt, and not only by the scientific establishment, but by the "enlightened" public at large—has in fact been vindicated through the discovery of quantum mechanics, which turns out not to be a mechanics at all. It appears that the physical universe—the universe as conceived by the physicist—cannot actually be separated from the interventions effected by the physicist himself; as John Wheeler has put it, we have been forced to admit that physics deals, finally, with "a participatory universe." What "breaks a physical system into parts," it turns out, is the empirical intervention by which the parts in question are specified; and because the measurement of one observable has an uncontrollable effect upon its so-called conjugate, it follows that the system as such can no longer be conceived as a sum of well-defined parts. This holds true, moreover, even for a system comprised of a single particle: for if one measures, say, the position of the particle, one inevitably disturbs its momentum, so that the system itself (in this instance, the particle) turns out to be inherently protean, a thing which is not, and cannot be, fully specified in mathematical terms. The notion of universal mechanism, championed by Galileo and seemingly confirmed beyond reasonable doubt by the discoveries of Newton and his successors, proves thus to be untenable. It turns out, a century later, that the Goethean denial was in fact well-founded after all: physics itself has now confirmed that conclusion, howbeit by a vastly different approach.[11]

The primary obstacle, which for so long had impeded our understanding of Goethe's scientific opus—namely, the hypothesis of mechanism, and more importantly still, the Cartesian philosophy upon which that premise was based—has thus been, in principle, overcome.[12] And yet that impediment remains with us as the quintessentially Cartesian conception of the "machine," which evidently constitutes the presiding paradigm of the technological society. It is neither a small nor a harmless thing to be surrounded on all sides by machinery, by "levers and screws" as Goethe says. In time, and by a kind of inexorable logic, the machine paradigm tends to impose itself within the technological society upon all aspects of human culture: our very conception of human society, and of man himself, tends finally to submit to its sway.[13] The result, needless to say, is a profound alienation from Nature: from the natural world around us to the "anthropic" world within. Either of these "worlds" has thus become, for us, a "closed book." We can, of course, theorize

11. It is no wonder that in the wake of this discovery there has been a surge of interest in Goethe's scientific opus, which in days gone by had been dismissed as the work of an amateur.

12. I say "in principle," because it happens that scientists, virtually without exception, are still imbued with the Cartesian assumptions. To be precise, I know of only two physicists who have transcended that philosophical premise, or have yet so much as recognized its hypothetical nature.

13. I have dealt with this question in *Cosmos and Transcendence*, op. cit., chap. 7.

about both, and we do so prodigiously; but we can no longer "enter," can no longer "see." Now, it goes without saying that this state of affairs closes the door to even the most rudimentary understanding of the Goethean doctrine. To so much as begin to comprehend his way of science, we need evidently to reverse the aforesaid "evolution" within ourselves: to recover, that is, a normal and authentically human relation to Nature, or better said, to all that lives and breathes within her.

We have noted that the Goethean science rests upon *"Anschauung"*: an intuitive penetration of the visible aspect presented by the phenomenon; we need also, however, to realize that such an "intuitive penetration" presupposes a profound kinship between man and Nature: the human microcosm and the cosmic macrocosm. The fact is that Goethe was deeply cognizant of that kinship: "if the eye were not sunlike"—*"wäre das Auge nicht sonnenhaft"*—he declares, "we could not behold the Sun." So too he sensed that Nature is something marvelous, something utterly profound and mysterious, which needs to be approached with a kind of reverence—again, the very opposite of the Baconian outlook, which regards Nature as something to be "harnessed for profit," as befits a machine. In what is obviously a rebuke of the Newtonians, Goethe declares the impotence of their empirical means: "What Nature does not freely disclose, you will not wrest from her with levers and with screws" (*"zwingst du ihr nicht mit Hebeln und mit Schrauben ab"*). Of course Goethe was cognizant of the fact that "levers and screws" have their use in the sphere of technology; what he denied is that such means can lead to a genuine knowledge of Nature: of that "which shows itself in itself."

But there is more: Goethe's science is based, not only upon a profound kinship with Nature, but also upon a deep *love*: a love which cannot but be near to what religion knows as "the love of God." If Nature be more than a mechanism—more than an inert machine—it must be something noble and beautiful and instinct with power; and that, to be sure, *is* something worthy to be loved. One senses an almost Franciscan quality in Goethe's relation to what he termed *"Natur."*

I would like now to point out that inasmuch as the object of

Goethean science does not reduce to a mechanism, nor to a sum of parts, but constitutes a veritable whole, that science does not deal, strictly speaking, with *quantities*: it cannot. Quantity, after all—as Aristotle has shrewdly observed—is "that which admits mutually external parts," which is precisely what the Goethean whole does *not* admit. One can say, once again, that "mutually external parts"—and hence, quantities—come "*later*." What exists at the level of the Goethean whole are not quantities, therefore, but *qualities*, precisely. And since the primary qualities pertaining to "the visible aspect" happen to be *colors*, it is not surprising that Goethe's scientific opus commences with his *Farbenlehre*, his "theory of color." What, then, *is* that Goethean science: what exactly does it accomplish? Strictly speaking, it deals, not with colors as such—which cannot actually be described—but with the conditions under which colors manifest, and which affect or determine that manifestation, something which *can* in fact be treated with scientific exactitude, and exhibits rigorous and previously unknown laws. The *Farbenlehre*, correctly understood, is in fact precise to the point of being in a sense "mathematical," without however, in any way, "quantizing" its subject, that is to say, "colors" properly so called. What gave rise, moreover, to Goethe's famous dispute with the Newtonians on the subject of "color" was not these findings, which are indeed scientific and which no one could deny, but the Newtonian contention that color can, in effect, be reduced to *quantity*—that is, to wave-length or frequency—a notion which Goethe adamantly opposed. That color was associated with a wave-length or frequency he did not deny; but he insisted that it has nonetheless its own reality, that in fact it "precedes" the quantitative parameters of the Newtonian conception. What Goethe rejected, as one can understand in retrospect, was not in fact the Newtonian physics as such, but the misbegotten Cartesian metaphysics, upon which that physics was, at the time, officially based. It seems that Goethe would not have disagreed with a Newtonian physics shorn of its metaphysical pretensions, a physics conceived strictly according to the Baconian recipe, that is to say, from an inherently operational or pragmatic point of view. It is only that Goethe would not have dignified such a discipline with the epithet "science"; he would most likely have subsumed it under the heading of "technology," the application of "levers and screws."

We should at least mention the second major field of Goethe's scientific endeavors, which is the so-called "metamorphosis" of plants. Given that the true object of Goethean science constitutes "a whole which is no-thing," his interest in plants is readily understandable: after all, a whole which is "no-thing" is perforce an *organismal* whole, of which the simplest and in a way most basic example is indeed the plant. Again, it would take us too far afield to speak of the Goethean "botany" even in summary fashion: like his *Farbenlehre*, the subject is demanding, and in its own way technical. Suffice it to note that either discipline springs from the Goethean *Anschauung*, to which it remains ancillary. To follow Goethe, it is therefore needful, in the final count, to acquire a corresponding "eye"—one that is truly "*sonnenhaft*"—and that is something not many, even among his most ardent followers, have apparently been able to accomplish.

The question arises now whether the Goethean *Anschauung*, which categorically exceeds our ordinary "seeing," is to be taken as "the last word" in the metaphysical quest. Having noted that the veritable *anthropos*—man in his integral reality—comprises not only a periphery, but indeed an absolute center, together with a hierarchy of intermediary centers, corresponding to so many distinct "levels of vision," one may ask to which of these "centers" the Goethean "vision" is to be assigned. Suffice it to say, in light of sapiential tradition—Christian and non-Christian alike—that the latter does *not*, by any means, constitute the innermost or *non plus ultra* "seeing": it cannot, because it constitutes yet a "creaturely" mode of perception. Let us attempt, now, to speak of the ultimate "seeing," which—strange to say—is in fact a seeing "with the Eye of God." It is here, and here alone, that the metaphysical quest—which, as we have noted, begins with the simplest act of sensory perception—reaches its term. We propose, now, to consider that "ultimate seeing" from a distinctly Christian point of view, based upon the teachings of Meister Eckhart, the controversial Dominican who did in fact affirm:

My eye and God's eye are one eye and one seeing, one knowing and one loving.[14]

We need to speak in the first place of the primary Center in man—the "center of centers"—which transcends not only what we term "body," but also "mind," even in the highest connotation of that term. Eckhart refers to it as the *vünkelin* or "little spark" in us, which he declares to be "*increatus et increabile*" ("uncreated and uncreatable"). It is apparent from the start that the Eckhartian anthropology transcends not only the customary "*corpus-anima*" conception of man, but the Goethean as well: where Goethe speaks of an eye that is "sunlike," Eckhart refers to one that is actually *divine*; and whereas the former is able to behold "the Sun," the latter, Eckhart assures us, beholds none other than *God himself*. Now, in view of that contention and all that it implies, it is hardly surprising that the Eckhartian teaching has been, from the start, a source of controversy, condemned by some, and hailed by others as the last word. Actually, what the Meister confides—in his "mystical moments," when he speaks, as it were, from the standpoint of God—is in fact what Scripture terms "strong meat" as opposed to "milk"[15]: such, in any case, is the premise on the basis of which we shall proceed.

But if *strong meat* "belongeth to them that are full of age," as the author of Hebrews declares, why then expound the doctrine in question in an essay addressed to all: to "young and old" alike? The reason, let me say, is that these are very special times: fearful times in fact. The Christian seeker finds himself hard pressed on all sides by the dominant currents of our day, which despite their often benign and indeed seductive appearance turn out to be anti-Christian to the core. There may be no more wild beasts in the Coliseum, but this obvious advantage is offset by the fact that the burning faith of bygone days and the spirit of brotherly love, which united

14. Sermon 12.
15. 1 Cor. 3:2, Heb. 5:14.

the early Christians and gave them immeasurable strength, have likewise disappeared. More than ever before, it is now up to the individual believer to stand his ground. There is however one compensating advantage which he enjoys: we have today access to the highest doctrines, teachings which in days gone by were available and permitted only to the few, perhaps in part because the many had no need for these high teachings. In our day, on the other hand, the need is there, and perhaps also a certain aptitude on the part of many, which previously was lacking: despite its progressive decline, to which we have already alluded, there is reason to believe that mankind may nonetheless be growing "older." Perhaps, for the earnest seeker of truth, the time for "strong meat" has now arrived: when it becomes almost miraculous to survive a college education without losing one's faith in God and religion, it appears that the times in fact *demand* as much. Now, it happens that the Eckhartian doctrine places into our hand a sharp weapon—a veritable "sword of gnosis"—which enables us, in principle, to "decapitate" spurious doctrines at a single stroke. Admittedly there is danger in this, and Clement of Alexandria is no doubt right in observing that "one does not reach a sword to a child"; but I surmise that even if one may not have "grown up" sufficiently to be fully trusted with such a "sword," the gain these days may be worth the risk. Besides, the very fact that someone opens this book is in itself an auspicious sign!

Let us then get back to the *vünkelin*, the divine spark hidden in the depths of our soul. Eckhart informs us that this "innermost" Center has itself a "structure": it is not "indecomposable" like a mathematical point, but comprises, formally speaking, two elements: a "ground," namely, and an "Image." We need not presently concern ourselves with the former, to which Eckhart refers metaphorically as "a vast wasteland" and "a solitary wilderness"; what needs to be considered is precisely the "Image." What, then, and "*of what*" is that Image? The answer to this question, as the reader may surmise, is that the Image proves to be none other than the Word or Son, who is indeed "the image of God" as St. Paul himself affirms.[16]

---

16. For instance, in 2 Cor. 4:4.

It is however to be noted that the Word does not stand alone, but pertains to the Holy Trinity, comprised of Father, Son, and Holy Spirit. And let us remind ourselves that this teaching constitutes in fact the central mystery of the Christian religion, a truth pondered and meditated upon by the Fathers and Doctors of the Church, which yet, in the final count, transcends what the human mind is able to fathom. To "conceptualize" the Trinity—to treat it as we treat other things—is thus already to miss the mark. The subject needs in fact to be approached "with folded hands"; and whosoever does not comprehend what this means—whosoever, in other words, lacks a "sense of the sacred"—will in no wise find access. It is easier, by far, to grasp the idea of the Absolute, or of the Unknown God, conceptions which in a way are native to the human mind, and have been upheld in all parts of the world since earliest times. The idea of the Trinity, on the other hand, pertains to Christianity alone, and is indeed inseparable from the Revelation bestowed upon mankind by Christ, the Incarnate Son of God. The fact, moreover, that this quintessentially Christian teaching is "rationally incomprehensible" proves to be of the utmost significance: it entails that the doctrine, when seriously upheld, may serve to activate within us a "more-than-rational" and indeed "more-than-human" faculty, which is none other than the Intellect, properly so called: a power, namely, which springs from the Image—from "the Christ in us"—and ultimately leads back into that Image.

To speak of the Trinity is to speak of the Divine Knowing: the Knowledge God has of Himself; and although that Knowing transcends, most assuredly, the divisions of time, and is said to take place in the *nunc stans* of eternity—in "the now that stands still"—it constitutes nonetheless a kind of "movement" and indeed a "life," paradoxical as this may seem. The ultimate "seeing"—the "seeing with the Eye of God"—is in fact none other than "life eternal" as Christ himself defines that life, in what theology knows as the "high-priestly prayer," spoken on the eve of his Passion: "And this is life eternal, that they might know thee, the only true God, and Jesus Christ whom thou hast sent."[17] It is clear that "the only true God" is

17. John 17:3.

indeed the Father; it needs also, however, to be understood that to see "the only true God"—to see the Father—and to see "Jesus Christ whom thou hast sent" are one and the same "seeing"; for "he that sees me sees him that sent me."[18]

It is thus by way of the *vünkelin*, the Word or Image "in us," that we are called to enter upon "life eternal"; as Christ declares: "I am the door."[19] How, then, does one "pass through that door": how "enter" into the Trinitarian life? This much one can say: it is a question of "seeing," of *gnosis* properly so called: it is by "sight" that one "enters in." But whereas all that has been said up to this point constitutes, in essence, the common teaching of Christianity, Meister Eckhart tells us more: not only does he say that "to see the Image" is in truth "life eternal," but he adds an absolutely startling declaration: that in fact *all* "seeing"—all "knowing" whatsoever—is ultimately a "seeing" of the very same Image! What makes the difference is the *way* we "see," the *kind* of "seeing" it is. Such is the stupendous claim which stands at the heart of the Eckhartian teaching, the epistemological masterstroke which holds the key to his entire doctrine. We must now consider this Eckhartian premise to the best of our ability.

We will base our exposition on one of Eckhart's German sermons,[20] a text which takes us into the very heart of the subject. It expounds the familiar words of Christ, normally translated to read: "A little while, and ye shall not see me; and again, a little while, and ye shall see me."[21] Eckhart, however, understands the "*modicum*" of the Vulgate text, not in a temporal sense—as "a little while"—but simply as "a little something," whatever it might be. And so he begins his sermon with the words: "However small a thing it is which sticks to the soul, we shall not see God." With this exegetical gambit— which apparently no one before him has ever conceived—Eckhart puts into our hand the key to metaphysics at large: everything is comprehended, ultimately, in this single magisterial declaration. Whether we perceive the cosmos, the things that pertain to what

18. John 12:45.
19. John 10:9.
20. Sermon 69.
21. John 16:16.

theology knows as "the order of creation," or perceive instead the eternal Word depends entirely on the condition of our soul: whether something "however small sticks to it" or not. Truly: "Blessed are the pure in heart, *for they shall see God.*"[22]

I will mention, in passing, that the Eckhartian teaching holds the key, not only to metaphysics (as we have said), but to *physics* as well: to the understanding, namely, of what we conceive to be *cosmic* reality.[23] What presently concerns us, however, is the fact that Eckhart lays bare, at the same time, the means by which the higher degrees of "seeing"—right up to the vision of God, the supreme *gnosis*—are to be attained; and let us note that this Eckhartian recognition constitutes in fact the principle behind all authentic yoga, be it of Eastern or of Western provenance. What, indeed, *is* yoga? Clearly, it is discipline which aims at the removal of the aforesaid *"modicum."* What, then, are those "little bits" that attach themselves to the soul, and in so doing obstruct our vision of the Word, the true Image? One cannot actually say what they are; for the *"modicum"* itself is never visible: it is not what we know, or *can* know. Patanjali, in the *Yoga Sutras*,[24] refers to these elusive "bits" as *chittavṛitti*, "modifications of mind," which is to say that they arise from mind (*"chitta"*), are carried by mind, and subside again into mind.[25] They are something, therefore, which has no essence, no existence of its own; like waves on the surface of the sea, they are nothing apart from the water. And yet, "non-existent" though they be, it is these "modifications" (*"vṛitti"*) that cause us to perceive "the ten thousand things" of this world in place of God: in place of that which *is.*[26] To put it in Biblical terms: it is they—these *chittavṛitti*—

---

22. Matt. 5:8.

23. In this context the Eckhartian principle is tantamount to what I term "anthropic realism," a position which proves to be crucial to all cosmology, and in particular, to the understanding of contemporary science, most especially quantum theory. See *Christian Gnosis: From St. Paul to Meister Eckhart* (Tacoma, WA: Angelico Press/Sophia Perennis, 2012), chap. 2.

24. The primary textbook on yoga, according to Hindu tradition.

25. One needs however to realize that "mind" by itself—mind "without modifications"—is no longer "mind" as we understand the term.

26. The reader may recall the *nomen Dei* of Exodus 3:14: *"Ego sum qui sum."*

that "make the heart fat," so that "in seeing," the people "perceive not."[27]

As Eckhart gives us to understand, these considerations entail a metaphysics which vastly transcends all our "dualistic" conceptions of God, man, and cosmos. Hidden doubtless in Scripture and in the words of the wise, it is in the Eckhartian opus—and especially in his German sermons—that this "secret" metaphysics comes at least partially to light. Eckhart's doctrine, as I have argued elsewhere,[28] is based upon the recognition that *knowing*—namely, *"seeing"*!—takes precedence over *being*. There are, ultimately, two modes of knowing—with or without "mental modifications," namely—which Eckhart identifies as the human and the divine. To know "without modifications"—without *"media,"* as Eckhart says—is to know as God himself knows; and what is thus known is the Word or Son of God, and through Him, God the Father. To know "with modifications," on the other hand, is to know in a creaturely way; and what the creature knows falls short of the mark—falls short of reality—which is and can be none other than the Word itself. To be precise: all that is known by way of a medium, whether it be a sensible image or a mental conception—everything, thus, that "is not God," is not divine, is not the Word itself—Eckhart terms "creature."

It should be noted that this knowing "through media" whereby we perceive the things of this world is indeed the "seeing" to which St. Paul refers in the famous dictum: "now we see as through a glass, darkly."[29] Clearly, what "darkens" our vision—like dust upon a "glass" or mirror—are precisely the "impurities" to which Eckhart alludes, which are said to "cling to the soul." There is also, however, a second mode of vision, and St. Paul refers to it without delay: having stated that "now we see as through a glass, darkly," he goes on to say: "but then face to face." The Pauline "then" stands thus in contrast to the "now," and refers evidently, not to a past or future moment in time, but to an alternative mode of knowing: a knowing "face to face," that is to say, *without medium*. But what is the nature

27. Isa. 6:9.
28. See *Christian Gnosis*, op. cit., chap. 6.
29. 1 Cor. 13:12.

of that second knowing? This question too the Apostle answers in the very same verse, which ends with the phrase: "even as also I am known." Now, the original text has "*epegnosten*" (which actually means "I *was* known"), an expression which refers specifically to the "supreme" knowing, what St. Paul terms "*epignosis*" as distinguished from "*gnosis*." What stands at issue is a knowing of God the Father conforming to the Christic definition of "life eternal"[30]: this is what it means to know "face to face." To "see *without medium*" is thus to see, not in a creaturely manner, but indeed "with the Eye of God" as Eckhart declares. One finds that in this single Pauline verse (i.e., 1 Cor. 13:12) the Eckhartian doctrine is in truth comprehended.

The metaphysical quest—which is none other than the task of religion according to its highest conception—reduces thus to a cleansing that rids the soul of its impurities: those intangible and elusive "little bits" that stick to it and impair our vision.[31] We are called to the very "purity of heart" by which we "shall see God." Nothing less than this will do: such is the perfection Christ has enjoined upon us[32]; and this is what Eckhart holds up unequivocally as the universal norm, the very definition of what he terms "the just man." Here is how he delineates that norm:

> I say in truth, as long as something takes form in you that is not the eternal Word and does not derive from the eternal Word, no matter how good it might be, that is really not right. Hence only he is a just man who has annihilated all created things and stands without distraction looking towards the eternal Word directly, and who is formed therein and is reformed in justice.[33]

We are told that the "just man" is one in whom "nothing takes form which is not the eternal Word": what does this mean? In light

30. Eph. 1:17–18.

31. The very first verse of the *Yoga Sutras*, in fact, defines yoga as *chittavṛitti-nirodha*, the "uprooting" (*nirodha*) of the mental modifications.

32. Most explicitly in Matt. 5:48: "Be ye therefore perfect, even as your Father which is in heaven is perfect."

33. Sermon 16b.

of the preceding considerations it can only mean that our vision is no longer impaired, no longer distorted by *media*. And that is the reason why the just man "has annihilated all created things": having "uprooted the modifications" he no longer beholds "creatures," but now sees in all things "the eternal Word" itself. Having "annihilated all created things" he literally "stands without distraction looking towards the eternal Word."

The great question, now, is how this Herculean feat is to be accomplished: how does a man "annihilate all created things"? And who can actually accomplish that? Eckhart replies to these questions in his sermon on the "*modicum*" text, which (as we have noted) he renders in the words: "A little bit, and ye shall not see me; and again, a little bit, and ye shall see me." And his answer is simple: it is the second "*modicum*," he declares, that destroys the first. But what *is* that second *modicum*, that second "little bit"? It is none other than what he elsewhere terms the "*vünkelin*," the "little spark" in the soul that is said to be *increatus et increabile*. This too is a *modicum*—a "*little bit*"—but of a very different kind. That second "little bit" Eckhart now identifies as the Word or Image in the soul; and that Image, he goes on to say, is the source of the power by which the "mental modifications"—the impurities of the soul—are to be subdued. And let us understand it well: this power is not human, is not "creaturely," but is—and needs perforce be—*divine*. It is in fact none other than the Holy Spirit "who shall lead you into all truth" as the Savior declares.[34] According to Eckhart's analysis, He does so by uprooting the "modifications"—what theology knows as "sin," or as the effects thereof—which prevent us from seeing the Word. That Holy Spirit, however, is "sent" by Christ; or as Eckhart has it: springs from the *vünkelin*, the Image which is the "Christ in us."

Let this suffice; enough has perhaps been said to provide at least an initial glimpse into the heart of the Eckhartian teaching, sufficient to indicate that "it is all about *seeing*": whether we see "as through a glass, darkly," or "face to face" to put it in the words of St. Paul. It is here, in this pivotal realization, that religion and metaphysics finally meet: that each recognizes itself in the other. And let us not fail to

34. John 16:13.

note that they meet thus in Christ, in Him who is "the way, the truth, and the life"[35]: the "way," because He cleanses and empowers our "eyes"; the "truth," because He is what "the pure in heart" shall see; and the "life," because thus to see God is indeed "life eternal."

It should be noted that the "modifications" which obstruct our vision—which prevent us from "seeing God"—divide into what the Vedanta terms "*kośas*" or "sheaths," which we may think of as so many "layers" or "shells," one within the other; and this means that there are, in principle, two ways of eliminating these obstructions: all at once—as happened, presumably, to St. Paul on the róad to Damascus—or "one by one," starting from the outermost *kosa* and proceeding, step by step, towards the innermost. And needless to say, it is the second of these options—what St. Bonaventure terms the "*itinerarium mentis in Deum*" or "mental journey into God"— that constitutes the "normal" way of spiritual ascent. The journey, however, is by no means "continuous," but proceeds, as it were, by "quantum jumps": from one "level" to the next. In this manner the *viator* passes successively through the various "intermediary" centers to which we have previously alluded; and to be sure, at each of these degrees he has the option of "lingering," and faces the danger, one might add, of falling back to a lower state.

Having touched upon the subject of "phenomenology" as practiced by its seminal representatives—Goethe and Husserl—one now sees that the phenomenological approach is indeed inherently yogic, a matter of eliminating (or circumventing) "modifications," in short, of "cleansing the mirror" by which we perceive. To see "earlier," as phenomenologists are wont to say—"before" the breakup into the empirical object and its subject has taken place—is to obviate the corresponding modifications, and in so doing, to realize a less mediated and consequently "higher" mode of vision. But whereas the phenomenological methodology can no doubt take the qualified practitioner a certain distance along the path of ascent,

35. John 14:6.

and may enable him to transcend, in some degree, the "blindness" that has befallen humanity at large, it is also clear, in light of sapiential tradition, that these means cannot take us "all the way." The phenomenological method, even at its best, does have its irremediable bounds, a fact which, apparently, Husserl himself discovered in his later years; as we have noted before, he came in the end to recognize his own inability: "I have been seeking God without God" he admits. And let us add that such is indeed the crucial recognition, the Socratic profession of incapacity which finally opens the door: to know "that we do not know"—and that, as the Savior declares, "Without me ye can do nothing"[36]—constitutes indeed the prime precondition to enlightenment.

We need to understand that the "blindness" of which Scripture apprises us—and which the methods or means of yoga, in the widest possible sense, are designed to cure—has been brought on, not simply by the primordial Fall, but by all the subsequent human betrayals, large and small, down through the course of history. It is therefore apparent, in light of the Judeo-Christian tradition, that "earlier," phenomenologically speaking, correlates with "earlier" in a historical sense. It emerges that what the phenomenological means enable us to accomplish, at least to some degree, is finally the retrieval of states corresponding to earlier periods, to an age in which mankind was less blinded than it is today. What these methods cannot do, on the other hand, is to "reverse the Fall": for this one needs the very "power" of which Meister Eckhart speaks, which is finally none other than the power of the Holy Spirit.

Authentic science seeks to grasp the phenomenon: "that which shows itself in itself"; all else is a "half knowing" at best. But what *is* "the phenomenon"? The answer to this question is given in the Eckhartian doctrine: in the recognition, namely, that *what is known* *"without medium"*—and thus "in itself"!—*is none other than the Word*. Start with whatever you will and seek "that which shows itself in itself": and in the end you will find the Word. You *must*: there *is* in truth nothing else to be found! The Word is the Only-Begotten Son of the Father who contains within Himself all that ever was or

36. John 15:5.

ever will be; as St. Paul affirms of the Incarnation: "In him resides all the fullness of the Godhead bodily."[37] It is not "mere poetry" when Christ declares to his disciples: "Inasmuch as you have done it unto the least of these my brethren, you have done it unto me."[38] A purportedly Christic logion, recorded in an apocryphal Gospel, epitomizes that teaching: "Split the wood, and you will find Me." Whether it be wood or stone or anything: if you penetrate to its core, its very essence, you will encounter Him. You must: because the essence of all things is contained in the Word. And that Word is a Magnet which draws all things unto itself, and *into* itself. In his "*modicum*" sermon—in a passage of rare beauty—Eckhart speaks of this supreme attraction and universal "destiny":

> You must understand that all creatures are by nature endeavoring to be like God. The heavens would not revolve unless they followed on the track of God or of his likeness. If God were not in all things, Nature would stop dead, not working and not wanting; for whether you like it or not, whether you know it or not, Nature fundamentally is seeking, though obscurely, and tending towards God. Nature's quarry is not meat nor drink...nor any things at all in which there is naught of God, but covertly she seeks and ever more hotly she pursues the trail of God therein.

We can do no better than close with the words in which the Meister himself concludes his Sermon: "To the end that we may grasp this and become eternally happy, may the Father and the Son and the Holy Spirit help us. Amen."

---

37. Col. 2:9. Regarding the significance of the adjective "bodily" ("*somatikos*" in the original) I refer to my treatise on Christian gnosis, op. cit., especially the chapter on Jakob Boehme.

38. Matt. 25:40.

ACKNOWLEDGMENTS

Chapters 1 through 7 are based upon articles which have appeared in the journals *Sophia* and *Sacred Web*. Chapter 8 constitutes an English version of a chapter in *Qu'est-ce que la metaphysique?* (Paris: L'Harmattan, 2010), edited by M. Bruno Bérard. The author wishes to thank the editors of *Sophia* and *Sacred Web*, as well as M. Bruno Bérard, for permission to republish.

INDEX OF NAMES

227

CPSIA information can be obtained
at www.ICGtesting.com
Printed in the USA
LVHW031225070720
659963LV00002B/159

9 781597 311359